ARThematics Plus:

Integrated Projects in
Math, Art and Beyond

A Reinforcement and Enrichment Resource for
Classroom Teachers, 4-6

by Stefanie Mandelbaum
and
Jacqueline S. Guttman

ISBN: 0-7596-2270-1 (e-book)
ISBN: 0-7596-2271-X (Paperback)

This book is printed on acid free paper.

1stBooks – rev. 02/12/03

Table of Contents

Introduction

Mathematics, on any level, is too often a mysterious, anxiety-provoking subject, even for many of the brightest students, and because math skills are cumulative, students who do not grasp the basic concepts in elementary school are doomed to struggle throughout their school years. These students, who avoid math courses whenever possible, not only miss out on the development of important problem-solving skills that will serve them in all facets of their lives, but, in an ambiguous world, never know the enormous satisfaction derived from finding the absolute *right* answer to a puzzle. Such students view mathematics as something they will never need in "real" life, not realizing that, in fact, the use of math skills is an integral part of daily living.

Interestingly, this notion of math as being "out there" is not unique to that subject. Art, too, is perceived as inconsequential to everyday needs. With slight variations (math is too hard, art is superfluous) students -- and sometimes their parents as well -- hide their intimidation and lack of understanding behind a smokescreen of dismissal.

Instead of viewing the world as an integrated whole, an ecological organism in the broadest sense of the word, we study and attempt to understand our environment by dividing it into what are often artificial compartments. The effect of this technique on many of our children has been to limit their ability to conceptualize - to see the larger picture. *ARThematics Plus: Integrated Projects in Math, Art and Beyond*, is an effort on the part of its authors to interweave these subject areas with one another and into the elementary school core curriculum as well. Each chapter also contains "tangents" -- extensions of the lesson into such areas as language arts, social studies, science and music.

Employing a multi-faceted, participatory approach, *ARThematics Plus* will help teachers engage students by demonstrating the use of fundamental mathematical concepts in the creation of works of art. While the primary goal is to reinforce these concepts, the authors hope as well that the book will enable students to become more comfortable with math and art, to see how both subjects fit into their lives, and convince students and teachers that neither solving math problems nor making art requires extraordinary gifts.

Our intention is to present information in a style that eschews academic jargon and esoterica in favor of accessible language that anyone can comprehend; in short, to provide a resource for both classroom teachers and math specialists. *ARThematics Plus: Integrated Projects in Math, Art and Beyond* is designed to help the elementary school teacher to comply with the latest NCTM Standards. The appendix contains a list of relevant websites.

ARThematics Plus:
Integrated Projects in Math, Art and Beyond

ARThematics Plus is divided into two major sections: <u>Shapes, Symbols and Symmetry</u> and <u>Pondering Proportions</u>. Each chapter consists of an overview, an art activity that reinforces a mathematical concept, follow up activities regarding the application of the lesson into the core curriculum and blackline masters of Ms. Mandelbaum's original illustrations. At the end of each chapter, there are definitions of words and terms as well as a list of additional resource materials.

Title	Mathematical Applications
SHAPES, SYMBOLS AND SYMMETRY	
The human need for balance is universal.	
2-D in a 3-D Universe	2- and 3-dimensional geometrical shapes; organic shapes; introduction to projective geometry; geometry in the world around us
Mystical Mandalas: Sacred circles from Chartres to Tibet	2-dimensional geometric shapes; four kinds of symmetry, organic shapes, patterns
Kitchen Floors and the Alhambra Palace: What do they have in common?	Tessellations, linear and angle measurement; multiplication, subtraction, division; calculators; symmetry; patterns; circles; translation, rotation, reflection, vertices; regular polygons
PONDERING PROPORTIONS	
Our ancestors' endless search for perfection.	
Collages and Composites	Prime and composite numbers; factoring; multiplication and division; fractions, decimals, percents
The Golden Mean: Plato, Fibonacci and Artistic Composition	Addition, subtraction, multiplication, division; fractions; patterns in number sequences; ratio and proportion; polygons; ellipses; calculators
Pythagorean Ratios in Art and Music	Addition, subtraction, multiplication, division; fractions, calculators, decimals, percents, weights and balance, patterns
Figurative Fractions	Addition, subtraction, multiplication, division of fractions; ratios and proportions, percents

Chapter 1
2-D IN A 3-D UNIVERSE

At a glance...

Time: One 90-minute session or two 45-minute sessions

Activity

An introduction to realistic and abstract artwork leads to a discussion of the geometric and organic shapes used by artists in their compositions. Students discover how artists draw what appear to be 3-dimensional geometric shapes and how these are projected onto a 2-dimensional surface. They see how the way they draw an object affects how real it looks. The relationship between linear perspective and projective geometry is explained. Tracing around 2-dimensional geometric shapes, members of the class create their own paintings. As they view and analyze one another's work, the students explain their choices and decisions.

Mathematical Concepts

♦ learning to recognize geometric shapes: 2-dimensional - circle, triangle, quadrilateral, rectangle, square, pentagon, hexagon, octagon, parallelogram, rhombus, trapezoid, polygon; 3-dimensional - cylinder, sphere, hemisphere, cube, pyramid, cone

♦ the difference between geometric and organic shapes

♦ discovering math all around us: geometric shapes in the classroom and on our own bodies

♦ introduction to projective geometry

Pre-planning

Read through the entire lesson ahead of time in order to know where it's going. Familiarize yourself with the glossaries of words and shapes. Decide whether you prefer to divide the lesson into two 45-minute sessions or complete it in a single session. For example, you might want to think in terms of a morning spent on math and art. Masters/transparencies should be duplicated ahead of time.

Materials/Equipment

1. Art work: 2-3 art posters or prints of similar subject matter (e.g. landscapes, still lifes, portraits) but different styles (e.g. realism, abstraction, surrealism, cubism, pointillism). While we have provided art that you can use for this lesson, if you have favorite paintings or drawings that you would prefer to use instead, please do so, first making certain that they contain a wide variety of geometric shapes.

*Note: The glossary at the end of this chapter defines these terms and gives examples of artists who work in each style.

2. Paper - preferably 18" x 24" if desk space permits
    We have found it to be most practical for students to work four to a table, sharing geometric shapes. Pushing four desks together also will serve this purpose.
3. 2-D and 3-D geometric shapes. Possible source: 2-D pattern blocks, 3-D wood geometric solids
4. Pencils with erasers
5. Rulers
6. Colored pencils, crayons, pastels, markers or paint and brushes
7. Transparencies made from our masters.
8. Overhead projector, if you are using the transparencies of art examples we have provided.

Step 1 - Art Introduction

Display the artwork you have selected (or use overhead transparencies provided).

For each piece of artwork, give the title, name of artist, nationality (by birth and residence) and when he or she lived. If you have other information available, share that as well. You may want to say a few words about what was going on in that part of the world when the artist lived there.

Venice Landscape (SM '98)   Venice Abstract I (SM '00)   Venice Abstract II (SM '98)

If you are using our drawings, try this introduction: Stefanie Mandelbaum drew these three landscapes.

Ask: What is a landscape?

A landscape is an outdoor scene. One of the above landscapes is realistic, which means it looks like what you would see in a photograph. The others are abstract versions of the same landscape.

Ask: Can they tell what makes it abstract?

Shapes have been simplified; details have been left out; some shapes have been changed completely; the lights and darks are in greater contrast. Overall, the landscape has been altered to become a more simplified pattern. Can they find the gondola, the bridge, the pole, the water, and the sky in all three of the Venice drawings?

Ask: What do these drawings or paintings have in common?

For example, they could all be:
    landscapes
    oil paintings
    painted by one artist or artists from the same part of the world

Most important for this lesson, all are made up of geometric and/or organic shapes.

Discuss the difference between geometric and organic shapes. (See glossary for definitions.) Remind the class that this discussion will focus on geometric shapes.

Step 2 - Math Introduction

Ask: What geometric shapes do we see around us in this room?

As the students identify them, discuss the difference between the 2-dimensional and 3-dimensional shapes they have found. Which are which? (You may wish to reproduce and distribute the Glossary of Shapes at the end of this chapter for your students' reference.)

Note: Certain other shapes, called **fractals,** are also the subject of mathematical study.

Examples:

2-D - map, charts and posters, clock face, floor/ceiling tiles, windows, pupil of eye

3-D - globe, cabinet, tissue box, paper towel roll, board eraser, chalk, trash basket

Point out: Artists have a special problem when trying to draw or paint a 3-D geometric shape, because the paper or canvas is in 2 dimensions and the shape is in 3 dimensions. They have to <u>project</u> the 3-D shape onto the 2-D surface. Let's see how they do it.

Look at some 3-dimensional shapes

Suggestions: Wooden geometric solids, coffee cans, ice cream cone, ball

    a.   Start with a coin, which we can think of as a "slice" of a cylinder. This can be demonstrated by stacking several coins of a single denomination. Each student can then look at his or her own coin. Holding it directly in front of one eye with the other eye closed, tilt the coin, making it gradually take on the appearance of different shapes:

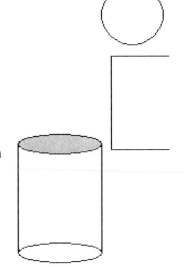

    b.   Hold up a cylinder in front of the class so that all that can be seen is the top or bottom.

Ask: When you see it this way, how would you draw it?
    A circle

    c.   Now hold it so the class <u>cannot</u> see the top or bottom at all.

Ask: What would you draw now?
    A rectangle, but it would require shading in order to appear to curve

    d.   Now hold the cylinder so that only a small portion of the top can be seen.

Ask: What would you draw this time?
    A rectangle topped by an ellipse

Repeat the same process with a cone, pyramid and cube.

Step 3 – Math-Art Discussion

Try creating a 3-D drawing with an everyday object like a tissue box. When we draw our tissue box, if we want it to be in perspective and appear to be 3-D, it should look like this:

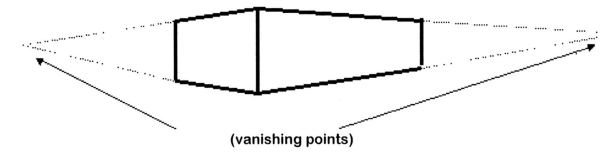

(vanishing points)

If you draw an object or building so that all the non-vertical lines must meet at a vanishing point (as shown), it makes what you are drawing look exactly right. By using a vanishing point, people can make accurate maps, charts, and scale drawings because it shows us exactly how much the lines should slant.

Another way to make objects appear 3-dimensional is by careful attention to their placement and size. Even though we may <u>know</u> that two objects are exactly the same size, our eye sees the ones that are farther away as smaller than those nearer to us. Here is an example:

Point out: There is a branch of math called "projective geometry" that deals with the problem of perspective -- creating the illusion of 3 dimensions on a 2-dimensional surface. Mathematicians think of projective geometry as a *visual* geometry because it deals with the way our eyes deceive us as we look at near and far objects and see them differently from different angles. Renaissance artists in the 1500's were very concerned with problems of perspective, but the actual mathematical theory came later. Projective geometry is an example of mathematicians learning from artists.

If you wish, tell the class this story about the origins of perspective and projective geometry.

The ancient Greeks knew how to use perspective in art, but for over a thousand years the achievements of their society were forgotten and neglected. Beginning in the early 1400's, however, people became very interested in the culture, art and architecture of the Greeks and Romans. This era is known as the Renaissance (meaning "rebirth"), partly because the laws and culture of ancient times were, in a sense, rediscovered. Early Renaissance artists like Giotto attempted to work with perspective but were not completely successful. If you were to look at their paintings, you would that see that the illusion of depth is not quite right because they did not use vanishing points.

Filippo Brunelleschi (broon-a-LES-kee) was a Renaissance architect and artist living in Florence. With his friend, the artist Donatello (don-ah-TELL-o), he took a trip to Rome to see the ancient ruins. As we might take photographs, Brunelleschi made sketches, and as he sketched, he realized that in order to make things look "right", he had to make the lines go a certain way.

He also observed that the more distant parts of the structures looked smaller than those nearer to him. He began developing his own rules about how to create the illusion of depth – what we would call linear perspective. One of his friends, Masaccio (ma-SA-cho), took Brunelleschi's ideas and used them in his paintings. Before long, other artists, including people like Michelangelo (mee-kell-AHN-gel-o), Raphael (RAH-fah-el) and Leonardo da Vinci (lay-o-NAHR-do da VIN-chi), started using these techniques as well, until it became the accepted and "correct" way to draw or paint.

In 1639, an architect and engineer named Desargues (da-SARG) entered the picture [so to speak]. He, too, had many artist friends. Noticing how they made use of perspective, Desargues realized that this was, in fact, a brand new kind of geometry, a geometry in which rectangles could turn into trapezoids and parallel lines could meet!

Draw these diagrams on the chalkboard to illustrate:

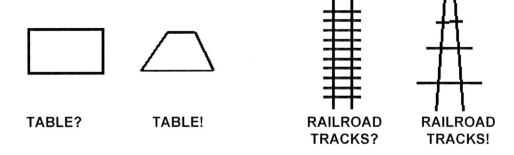

| TABLE? | TABLE! | RAILROAD TRACKS? | RAILROAD TRACKS! |

Interestingly, Desargues' ideas about projective geometry were ignored by mathematicians of his era, perhaps because they were so far ahead of their time that people were not ready to absorb and understand them. It was not until the 1800's that this concept was developed further and became a respected branch of mathematics.

An experiment for your students: Go to the window; draw a rectangle, or "frame" (or use a window pane); close one eye (so everything looks flat, or 2-dimensional). What do you notice? The objects in the background will always appear smaller. If you wish, place a piece of transparency paper on the window and have the students draw what they see. Distant objects will be smaller than those nearby.

Step 4 – Reinforcement

Look at the paintings again. Have the students identify the geometric shapes used by the artist(s) and tell what each represents. For example, in *Venice Landscape*, the windows are quadrilaterals, some of which are rectangles. Part of the bridge is a decagon (a 10-sided polygon). *Venice Abstract I* is composed of positive and negative shapes, some of which are organic and some of which are combinations of geometric shapes. In *Venice Abstract II*, there are various intersecting geometric shapes, such as triangles, trapezoids, pentagons, and hexagons.

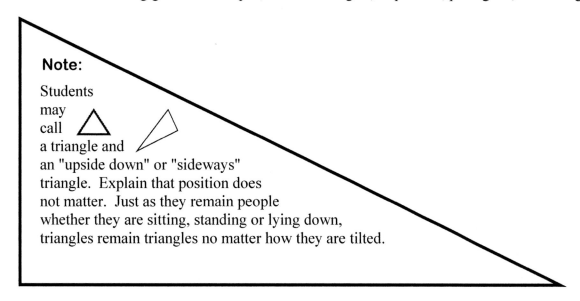

Note:

Students may call  a triangle and  an "upside down" or "sideways" triangle. Explain that position does not matter. Just as they remain people whether they are sitting, standing or lying down, triangles remain triangles no matter how they are tilted.

Point out: Artists often "play" with shapes to get the effect they are looking for, using them in different ways.

Ask: What kinds of different feelings or images do we get from the way artists use shapes?

Example: Jagged triangles might convey a scary feeling of being alone on a mountain at night, especially if used with dark colors or shades as we did here.

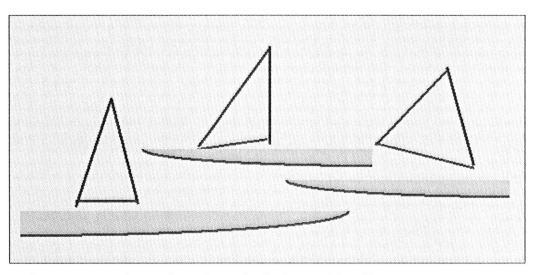

Used another way, triangles might make us think of peaceful sailboats.

If your students are familiar with equilateral, isosceles and right triangles, have them identify them as such when looking at the paintings.

Review Questions:

♦ Do the artworks make you <u>feel</u> a certain way? How are different feelings conveyed?

♦ Do the abstract landscapes <u>feel</u> different from the realistic one? In what way?

If using the Mandelbaum artwork, **point out** that the cylindrical pole is drawn almost as a rectangle and then shaded to give the illusion of 3 dimensions. Extend the non-vertical lines of the building on the right (windows, water line, etc.) to show that they meet at the same point (the vanishing point) on the left.

BREAK THE LESSON HERE

Tell the class that next time they will make their own art work that combines geometric and organic shapes.

2-D 3-D continued...

Step 5 - Art-Math Activity

Students make their own artwork using geometric and organic shapes.

Distribute paper, rulers, geometric shapes, and something to color with.

Suggest that they first think of the feeling they want, and then what kinds of shapes and colors they would use to convey that feeling. While it is not necessary that the students use a vanishing point, do encourage them to develop an awareness of how:

a. 3-D shapes are drawn differently depending on our angle of vision

b. objects in the distance are drawn smaller

View one another's work.

Students can either show their work, pointing out the use of both geometric and organic shapes and how they created 3-D effects or, alternatively, you can ask your students to find and identify the shapes in each other's work. If you wish to pursue the art aspect further, include a discussion of how they used both shapes and color to communicate what they were feeling or thinking about.

Note: Many combinations of paintings work equally well for this lesson. Some suggestions:

Landscapes

El Greco: *View of Toledo* (realism)
Salvador Dali: *Persistence of Memory* (surrealism)
Paul Signac: *The Harbor of St.Tropez* (pointillism)
Paul Klee: *Landscape* (abstraction)

Still Life

Georges Braque: *Le Jour* (cubism)
Paul Cezanne: *Still Life with Peppermint Bottle* (post-impressionism)

Portraits

Francisco Jose de Goya y Lucientes (Goya): *Don Manuel Osorio* (realism)
Paul Cezanne: *The Card Players* (post-impressionism)
Pablo Picasso: *The Three Musicians* (cubism)

Follow-up Art Activities

1. Cut geometric shapes (and organic shapes, if desired) out of colored paper and glue them together on a sheet of construction paper to make a "shape collage."
2. Go through magazine illustrations and cut out as many geometric shapes as you can find. Again, make a collage.
3. Draw a picture as realistically as possible. (A photo may be used instead.) Then try it in an alternative style, e.g. surrealist or abstract.
4. Look at various paintings from two different aspects: geometry and feelings. Artists communicate their feelings in many ways. Shape is one; color is another. After finding the shapes in a painting, you might also ask: What kinds of colors does the artist use? How does

the painting make you feel? Would you feel different if the colors were different? The paintings cited below can be found on the Metropolitan Museum of Art website given in the appendix. All of these artists were considered Spanish, although not all were born in Spain.

a. El Greco (1547-1614): *View of Toledo* - Note that most of the geometric shapes are in the architecture, e.g. parallelograms, trapezoids, triangular roofs. El Greco wanted to create a scary mood, so he used dark blues and greens, adding small amounts of white and light colors for contrast. The light colors make the dark ones seem even darker and scarier. El Greco lived during the time of the Inquisition. Do you think this affected the mood of the painting? (Mentioned in social studies section as well.)

b. Goya (1746-1828): *Don Manuel Osorio* - Point out the location of the geometric shapes (cage, paper, eyes, underlying pentagonal shape of the boy's head). Goya's painting seems simple at first, but as we look further there are ominous undertones. Is the cat on the left about to eat the bird? Is the boy sad, confused? The caged birds look almost dead; the bird outside looks alive but is in danger from the cat. Also, all the colors in the painting are dull except for the red, which makes you notice the boy -- and his wealth -- first. It is meant to distract you from the disturbing things lurking beneath the surface. Read about Goya's life. How do you think it might have affected his painting?

c. Dali (1904-1989): *Persistence of Memory* - The geometric shapes are on the left - trapezoid, parallelogram, ellipse, among them. Dali does a strange thing: he takes a geometric shape (circular clock) and, imagining that it is "melted", changes it to an organic shape. Why do you think Dali did this? Would it feel or mean something different if it were just a circle instead of a clock? Dali is more playful than El Greco-- not scary, just weird. How are his colors different from El Greco's? He uses warmer colors of gold, red-orange and yellow.

d. Pablo Picasso (1881-1973): *Three Musicians* - What do you see in the painting? What kind of instruments? What kind of feeling do you get? What might the music sound like if we could hear it - smooth? slow? jazzy? Note that not only the table, which is naturally geometric, is painted to *appear* geometric; the people are also broken down into geometric shapes. Picasso does the opposite of what Dali does with the clocks: Dali made geometric shapes (round clocks) into organic ones, while Picasso made organic shapes (humans) into geometric ones. How are the colors different from the other paintings? The brightest colors make you see the center figure first; the white on the right and left make you notice the other players; the dog in the background we notice last because our eyes usually see brighter colors before the brown and black of the dog. You might also point out that the painting is made to look like a collage.

Tangents: Related Activities for the Core Curriculum
Language Arts
1. Students can create oral or written stories about the artwork they have created. It is often easier to write or speak about something you can see.
2. Students can read *Flatland: A Romance of Many Dimensions,* a book about a 2-dimensional creature trying to comprehend three dimensions. You may prefer to read it aloud over a period of days or weeks.
3. Students can write imaginative essays or stories about themselves as 3-dimensional beings attempting to function in a 2-dimensional universe. Taking this further, what if they were to imagine themselves walking into a painting? Suppose it were an abstract 2-D universe? What if all the figures and objects were cubist? What if everything were surrealistic? Some examples of interesting 2-D "painted universes" would be:

Salvador Dali: *Persistence of Memory* (Surrealism)
Georges Braque: *Houses at Estaques* (Cubism)
Paul Klee: *Landscape* (Abstract)
Stuart Davis: *House and Street* (Abstract)
Edward Hopper: *Nighthawks* (Realism)
Wassily Kandinsky: *Composition No. 8* (Geometric Abstraction)

Social Studies
1. Look at paintings by artists from different countries and eras. Discuss the extent to which the society in which the artist lived and worked affected (or did not affect) his or her subject matter and style of painting. For example, do the students feel that Hopper's *Nighthawks* could have been painted in 15th century Europe? If Goya were painting a young American boy, what would look different? What might remain the same? Would *Toledo* have looked the same if El Greco had not been living at the time of the Spanish Inquisition?
2. Other recommended artists: Diego Rivera, Frida Kahlo, Georgia O'Keefe, Louise Nevelson, Romare Bearden, Jacob Lawrence.
3. Learn more about the Renaissance. Can the students find any aspect of Renaissance society that remains a part of our life today? (e.g. inventions, laws, architecture, literature, art, music and musical instruments) See Science section for more about this.

Science/Technology/Math
1. Study more about Brunelleschi, designer of the dome of the Cathedral of Florence, a very important example of Renaissance architecture. Construction of the cathedral began in the year 1296, during the Middle Ages, but when it came time to add the dome, the hole in the roof was so large (140 feet across) that the architects could not figure out how to build it so it would not collapse. Using his knowledge of Gothic building principals, Brunelleschi realized that, instead of trying to build an enormous hemisphere, they could use a series of Gothic

vaults, or arches, bending them inward toward a small center structure which would hold them together. It took from 1420 to 1436 to finish the dome. Brunelleschi also invented much of the machinery that was used to build it.

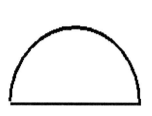

Hemisphere   *Brunelleschi's dome (diag.)*   *Il Duomo with Brunelleschi's dome*

2. In order to give your students a sense of the scope of Brunelleschi's achievement, have them measure its approximate size of 140 feet. One way to do this is by having a group of students cluster at a central point and walk off half the distance in several directions, like the spokes of a wheel. Imagine the challenge of building a dome that large! If they are learning about radius and diameter, this exercise might prove very illuminating.

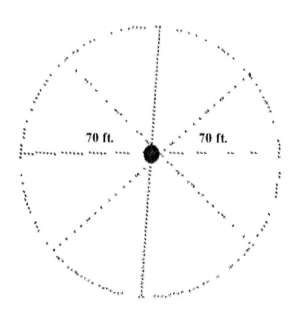

3. With further regard to the legacy of the Renaissance, find out some important scientific facts that emerged during that period. Examples:
    a. The Greeks knew that the world is round, a fact that was forgotten and rediscovered during the Renaissance.
    b. The classic scientific method of hypothesis and observation, developed by Leonardo da Vinci, is still in use today.

Movement

1. Students can use their bodies to create geometric shapes (more or less):

On the floor - make all kinds of triangles (e.g. 3 students of the same height, or 2 tall and one shorter) and identify them, if possible. Have the class select students to form a square (all the same height). They can also create a rectangle, pentagon, hexagon, and so on. Younger children may wish to "walk" or "skip" the shapes.

Standing up - ask them to arrange themselves into various types of triangles (a good teamwork exercise) and identify their own or those of other groups.

Large groups of students can "walk" various giant shapes across a gym or playground, much as a marching band forms different shapes. This is more fun with march music accompanying them. Hint: Decide on the shapes ahead of time; mark them with chalk, rope or tape.

2. Study Brunelleschi's dome. Have the students form such a dome by standing in a circle with their arms extended overhead and bent inward. Notice the cupola. Figure out what kind of round object (e.g. a small round wastepaper basket) they might all hold onto to form such a cupola. Take this a step further by choreographing a simple dance based upon the shape of the dome. Some ideas: open and close the dome; have the sections move outward and in a circle; have the sections break apart, move separately and rejoin one another. Add music that the students select.

For Further Reference

Edwin A. Abbott, *Flatland: A Romance of Many Dimensions,* Dover Thrift Edition

Kandinsky: Russian and Bauhaus Years, The Solomon R. Guggenheim Foundation, New York, 1983

Cynthia Goodman, *Hans Hofmann,* Abbeville Press, New York, 1986

Carolyn Lanchner, ed., *Paul Klee*, The Museum of Modern Art, New York, 1987

Georges Braque: The Late Paintings 1940-1963, The Phillips Collection, 1983

Luis Romero, *Dali,* Chartwell Books Inc., Secaucus, NJ, 1975

Glossary

Abstraction - refers to art which is nonrepresentational, stressing formal relationships between shapes, values, and colors (see Value) (Mondrian, Kandinsky).

Collage - (coll-AJH) a work of art created by pasting various materials on a 2 dimensional surface.

Cubism - art which reduces natural forms to their geometric equivalents (Picasso, Braque).

Fractal - a shape or structure that has detail ("bumpy" edges) no matter how large or small.

Geometric shapes - 2 dimensional shapes formed either by straight lines or by regular curved lines.

Geometric solids - 3 dimensional shapes formed by the combination of 2 dimensional geometrical shapes and/or regular curved surfaces.

Impressionism - a style of painting developed in the last third of the 19th century, characterized by short brush strokes of bright colors in immediate juxtaposition to represent the effect of light on objects (Renoir, Monet).

Landscape - a 2 dimensional artwork in which natural scenery is depicted.

Linear perspective - a technique which creates the illusion of 3 dimensions on a 2 dimensional surface (see Vanishing point).

Organic shapes - irregular shapes which suggest forms found in nature.

Perpendicular - meeting a line or surface at right angles

Pointillism - a technique based on the scientific discovery that small areas of pure colors placed close together are optically mixed by the eye (e.g. a section painted with blue and yellow dots will appear to be green) (Seurat, Signac).

Post-Impressionism - a varied development of Impressionism stressing formal structure (Cezanne, Seurat) or the expressive possibilities of form and color (Van Gogh, Gauguin).

Projective geometry - a branch of mathematics that deals with problems of perspective (see Linear Perspective).

Portrait - artwork in which the main subject matter is a person or people.

Realism - artwork which corresponds to ordinary visual experience, somewhat like a photograph (Hopper, Leonardo da Vinci).

Still life - artwork in which the main subject matter consists of stationary objects.

Surrealism - artwork which stresses nonrational, dream-like imagery (Dali, Magritte).

Value - the degree of lightness or darkness of a color.

Vanishing point - that point at which receding parallel lines appear to meet.

2-D GEOMETRIC SHAPES (Flat)

Polygon - *any* many-sided shape

Octagon - an 8-sided polygon

Hexagon - a 6-sided polygon

Pentagon - a 5-sided polygon; the building of that name in Washington, D.C., seen from above, has a classic pentagon shape

Quadrilateral - any 4-sided polygon

Parallelogram - a quadrilateral whose opposite sides are equal and parallel

Rectangle - a parallelogram in which all angles equal 90o

Square - a rectangle in which all sides are of equal length

Rhombus - a parallelogram with 4 equal sides. (A square is a kind of rhombus.)

Trapezoid - a quadrilateral in which only 2 sides are parallel

Triangle - a 3-sided polygon

Circle - a closed curved line on which all points are the same distance from the center point

3-D GEOMETRIC SOLIDS (Having depth)

Cylinder - A solid bounded by 2 parallel circles which are connected at every point by lines that are perpendicular to the circles

Cube - a solid bounded by 6 equal squares which are either parallel or perpendicular to each other

Cone - a solid bounded by a circle and a point outside the circle which is connected by a straight line to every point on the circumference of the circle

Sphere - a solid body whose surface is at all points at an equal distance from its center

Hemisphere - half a sphere

Pyramid - a solid having a polygonal base and triangular sides which meet in a point

Venice Landscape

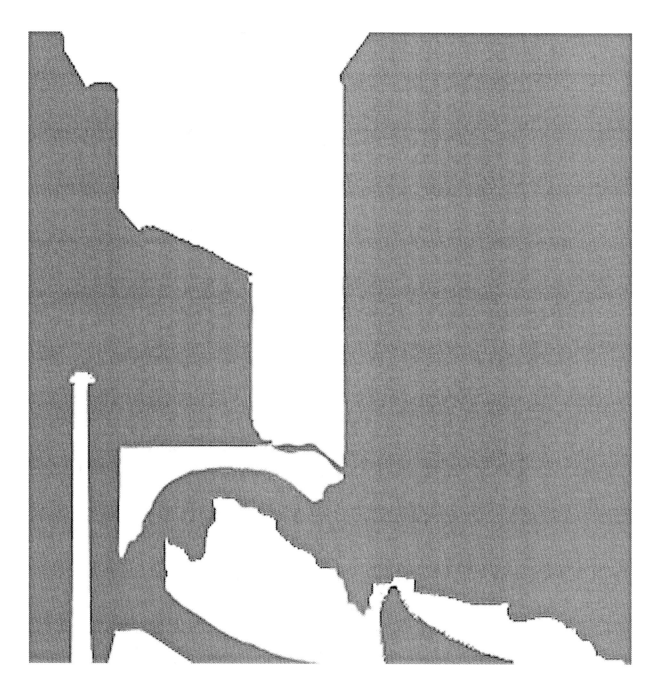

Venice Abstract I

Venice Abstract II

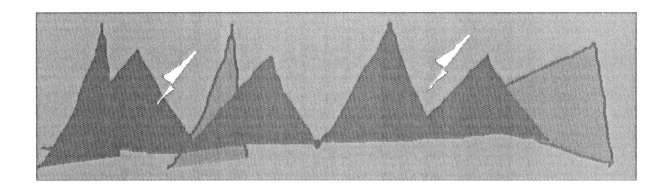

Chapter 2:
MYSTICAL MANDALAS

At a glance...

Time: Two hours

Activity

An illustrated introduction to the uses of mandalas in various cultures leads to a discussion of bilateral, rotational, expansion, and translational symmetry, as well as asymmetry. Using what they have learned about this concept, students create their own mandalas, incorporating the use of both geometric and organic shapes. Depending on the level and capabilities of the class, they may simply view each others' work and identify the prevailing types of symmetry, or learn about such concepts as *associative law*, *additive identity*, and *group theory*. Simultaneously, students are encouraged to develop an awareness of the universality of the human experience.

Mathematical Concepts

- ♦ review of geometric shapes
- ♦ recognition of different types of symmetry that exist all around us
- ♦ review of the meanings of diameter, perpendicular, right angle, congruent

Pre-planning

Read through the entire lesson ahead of time in order to know where it's going. Familiarize yourself with the glossary. Masters/transparencies should be duplicated ahead of time.

Materials/Equipment

1. Drawing paper
2. Compasses or, alternatively, large circular paper plates to trace around
3. Markers, crayons, colored pencils, paints or pastels
4. Geometric pattern shapes
5. Rulers
6. Transparencies made from our masters
7. Overhead projector

Step 1 - Introduction

Show mandalas that have been provided.

Mandala is a Sanskrit word meaning "sacred circle". Most mandalas contain one or more types of symmetry. Symmetry refers to repeated forms which follow systems of rules. Whether or not they are called mandalas, symmetrically designed circles are found in every culture and most time periods. Mandalas are often used to symbolize the universe as well as the inner self.

Mandala based on Bini mask from Nigeria.
Bilateral symmetry. SM '96

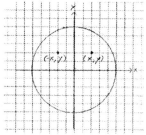

x-y axes showing bilateral symmetry

Underlying construction of Gothic rose window
Rotational and expansion symmetry. SM '96

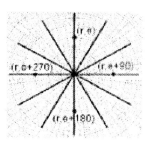

Polar coordinates showing rotational
symmetry

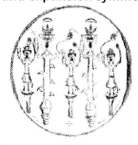

Mandala based on Navajo sand painting.
Translational symmetry. SM '98

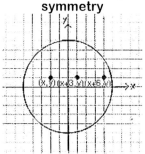

x-y axes showing translational symmetry

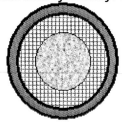

Mandala based on target shape. JG '00

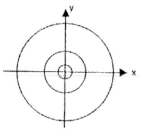

x-y axes showing expansion symmetry

Explain the derivation of each mandala as described above.

Ask: What do they have in common?

All are circles; each contains one type of symmetry.

Step 2 - Define the four basic kinds of symmetry

As you go through this process, have the students identify which type of symmetry is illustrated in each sample mandala.

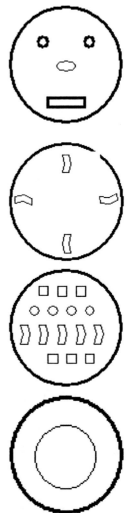

a. <u>Bilateral</u> (also called mirror symmetry) - A central line, or axis, divides the space so that the arrangement on either side reflects the other. The line may be visible or invisible. Examples: humans (faces and bodies), animals, even a chalkboard bisected by a line.

b. <u>Rotational</u> - The same shape is repeated at regular intervals around a circular area. Instead of a center line, there is a center *point*, visible or invisible. Examples: a clock, bicycle wheels.

c. <u>Translational</u> (also called wallpaper symmetry) – A shape moves across an area in a straight line in any direction. Examples: stripes or stars on the American flag, floor or ceiling tiles.

d. <u>Expansion</u> - Similar shapes of different sizes are "nested" inside one another surround-ing a common center, much like the effect of tossing a pebble in water. Examples: bullseye, bottoms of coffee cans.

Look at the additional sample mandalas provided.

Ask: Which types of symmetry do they illustrate? Are any of them asymmetric?

Some are asymmetric, that is, not symmetric. Others are symmetric, but contain asymmetric shapes or designs within the underlying geometric shapes.

Step 3 - Create your own mandalas

a. Using a paper plate, compass, or string and pencil, draw a circle. Cut it out.

b. Find the center of the circle. (This can be done by folding the circle in half and in half again.)

c. Divide the circle into quarters, either by using the folds already made or by lightly drawing two perpendicular diameters.

   Note: This is a good opportunity to teach or review the meanings of perpendicular, diameter, and right angle.

d. Using the geometric pattern pieces, coins or a ruler, create a basic symmetric design in pencil.

e. Add whatever kinds of organic shapes you wish. These may or may not be placed symmetrically and may or may not be congruent shapes.

f. The diameters may be erased or used as part of the design.

g. The designs may be colored however the students wish.

Alternative Method

Instead of measuring, students can estimate ("eyeball") the placements of the shapes in order to maintain the symmetry. In this method, the circle need not be cut out. Students might wish to draw outside the circle as well, maintaining at least one type of symmetry as they do so.

Variations: One-and-one-half hours.

♦ Use additional materials, e.g. colored paper, geometric shapes cut out of magazines, bits of cloth, colored pipe cleaners, etc. to make mandala collages.

♦ Create "permanent" sand paintings: Colors can be added to sand or corn meal using vegetable dye or liquid tempera.

a. Draw the mandala on a board.

b. Select a sand color.

c. Put glue on all the areas that will use the selected color; sprinkle the colored sand on those areas. Repeat the process with each color.

d. Glitter can be used to add sparkle.

e. When the glue is dry, the excess sand may be poured or gently vacuumed off the board.

♦ Use stencils, such as French curves, to make organic shapes that will maintain the symmetry.

♦ Draw and cut out organic shapes and use them instead of geometric shapes to create symmetrical mandalas.

Follow-up Math Activities

<u>Basic</u>

When the students have finished their mandalas, they should take turns showing them to the class. The other students should find and identify the type of symmetry used in each mandala.

<u>Advanced</u>

(Suggested for math-oriented teachers to use with gifted classes)

Discuss the properties of the *integers* under the operation of addition. They satisfy the *associative law*, they have an *additive identity*, they are *closed* under addition, and each integer has an *inverse*. The integers therefore form a *group*. Since they also satisfy the *commutative* law, they form a commutative or *Abelian* group.

Once the students understand what formulates a group, they can learn about symmetry groups using symmetric mandalas as an example.

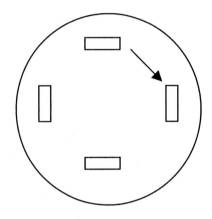

The inner circle is a mandala with <u>rotational symmetry.</u> The first element of this group, <u>Rotation no. 1</u> (R1), rotates the circle 90 degrees clockwise (illustrated). The second element of the group, R2, rotates the circle 180 degrees clockwise. The third element of the group, R3, rotates the circle 270 degrees clockwise. The fourth element of the group, R4, rotates the circle 360 degrees clockwise. R4 is therefore the <u>identity</u> of the group, since it brings the rectangle back to the starting position

Students can then check to make sure each element has an <u>inverse</u>. For example, the inverse of R1 is R3 because R1 + R3 = R4. Put into words, rotating the circle 90 degrees clockwise, followed by a rotation 270 degrees clockwise, is the same as a rotation 360 degrees clockwise.

They can also check to make sure the group satisfies the <u>associative law</u>. For example, (R1 + R3) + R2 = R1+ (R3 + R2). In words, rotating 90 degrees clockwise, followed by rotating 270 degrees clockwise, followed by rotating 180 degrees clockwise, is the same as rotating 270 degrees clockwise, followed by rotating 180 degrees clockwise, followed by rotating 90 degrees clockwise. These can be illustrated to be true by drawing the above mandala on the chalkboard and noting where the original rectangle is after each rotation. Students can check the <u>commutative law</u> to see if this is an <u>Abelian group</u>, i.e., does R1 + R2 = R2 + R1? Does a rotation of 90 degrees clockwise followed by a rotation of 180 degrees clockwise equal a rotation of 180 degrees clockwise followed by a rotation of 90 degrees clockwise?

A similar investigation could be made using any of the symmetric mandalas created by the students. Do their rotations always form a group?

Tangents: Related Activities for the Core Curriculum

After investigating the mathematical properties inherent in symmetrical mandalas, you and your students can travel down the many paths opened to us by the study of these sacred circles. Not only do mandalas in various incarnations appear in many otherwise dissimilar cultures, but the need for symmetry and balance, too, is endemic to most peoples.

Language Arts

After students construct their personal mandalas, they can follow this with individual or group stories about them, in either oral or written form. In addition, a discussion can be held about the use of symmetry in language. One type takes the form of a palindrome, e.g. madam I'm adam. (Even in this simple example, however, the spaces between the words make this phrase visually asymmetrical.) It would be an interesting exercise for students to try to create verbal palindromes of their own.

Certain types of poetry, too, can be symmetric. A line that recurs at the end of each verse, for example, is a form of translational symmetry. One famous example is Edgar Allan Poe's "Quoth the raven, 'Nevermore.'" from "The Raven." Students can read examples and try their hand at writing a poem using this kind of recurring refrain.

Social Studies

Students can research the use of sacred circles in different cultures. This may be related to any culture the class is studying, as mandalas appear in some form throughout the world.

In Tibet, elaborately and exquisitely decorated mandalas are made from colored sand which, when complete, are swept up and disposed of as part of a ritual. The process is a cross between visual and "performance" art, since the actual process of creating the mandala, sweeping it up and then pouring it out is more important than the product itself. The entire ritual is meant to help bring about world harmony and peace. In Tibet, the mandala is also used to assist meditation and concentration.

This emphasis on process is also found in Navajo sand paintings. In that civilization, a person who is mentally, emotionally or physically ill is considered to be out of harmony with the universe. The ill person sits outdoors throughout the night within a sand mandala which is meant to dissipate with the night wind, bringing the person back into harmony once again. Both the Tibetan and Navajo versions are concerned with healing -- the Tibetan with healing the world and the Navajo with healing the individual.

The use of sand in these two instances can provide a point of departure for a discussion of the relative importance of permanence and impermanence in different cultures.

Science

Using a free-standing mirror, students can study the bilateral symmetry of reflections. If they stand at one edge of the mirror so that only half a face, one leg, and one arm are visible, by moving the visible arm and leg up and down, they can create an illusion of floating on air!

Kaleidoscopes also use reflecting surfaces to create constantly changing, perfectly symmetrical designs. Students can investigate how they work, perhaps by "dissecting" one.

Music

Rounds or canons, in which the same melody is started over and over at specific intervals, are a form of translational symmetry. Try singing *Frere Jacques* or *Row, Row Your Boat*.

Row row row your boat gently down the stream...
      Row row row your boat gently down the stream...
            Row row row your boat gently down the stream...
                  Row row row your boat gently down the
                  stream...

Study the symmetry of *rondo* form, in which a melody is heard, followed by a new melody, followed by the first, followed by still another new melody, and so on. (ABACADAEA, etc.) This can be perceived as rotational, if invisible, symmetry.

Dance

Circle dances, in which participants perform symmetrically based movements, are used in many cultures, sometimes for fun and sometimes for sacred rituals. Perhaps a student who knows one of these dances can teach it to the rest of the class. Some examples: British/American - Hokey-pokey; Native American - ritual dances for a variety of purposes; Israeli – Hora; Greek - Mizralu; Italian - Tarantella.

Culinary Arts

Some students may benefit most from the tactile (and tasty) experience of making pizza or cookie mandalas, decorating them with different types of symmetric designs made from things like pepperoni, onion rings, peppers or chocolate chips.

Visual Art (Architecture)

While all kinds of symmetry are used in art, it is also great fun to find it in architecture. One kind that appears on many buildings in cities all over most of the world is in the decorative designs that may run across a roof line or along a wall. Even the windows themselves are often of a regular shape that repeats over and over horizontally or vertically. This, as we know, is translational symmetry. How many buildings can students find in their neighborhood or town that use translational symmetry?

In the 1700's, known in art and music as the "classical" era or period, form and balance became extremely important in architecture. Classical architecture often looked something like this:

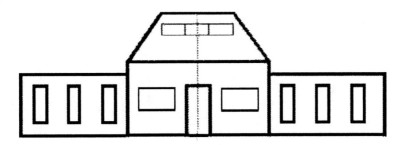

The class should be able to identify this building as having bilateral symmetry (note dotted line). Each "wing" of the building also has translational symmetry. Classically designed buildings are still very commonly used. Students can identify symmetrically constructed homes, schools, office buildings or official structures in their town or neighborhood, and even design or construct models of their own symmetrically shaped buildings.

For Further Reference

Examples of artists who have used mandalas in their work:

Kenneth Noland: *Warm Sound in a Gray Field*

M. C. Escher: *Circle Limit III*

Henri Matisse: *Design for Rose Window*

Kazimir Malevich: *Suprematist Element*

Fritz Glarner: *Relational Painting, Tondo 36*

Folk art/crafts that use mandalas: Amish quilts, Tibetan and Navajo sand paintings, Central American decorated wagon wheels, Gothic-style rose windows (on churches and synagogues).

Peter Gold, *Navajo and Tibetan Sacred Wisdom: The Circle of the Spirit*, Inner Traditions, Rochester, Vermont, 1994

Elizabeth ten Grotenhuis, *Japanese Mandalas, Representations of Sacred Geometry*, University of Hawaii Press, Honolulu, 1999

Philip Koslow, *Benin, Lords of the River*, Chelsea House Publishers, New York and Philadelphia, 1996

Fiona Macdonald, *Ancient African Town*, Franklin Watts, a division of Grolier Publishing, New York, London, Hong Kong, Sydney, Danbury, 1998

Christopher Wilson, *The Gothic Cathedral*, Thames and Hudson Ltd., London, 1990

Glossary

Abelian (uh-BEEL-yan) (group) a commutative group (see below) named after Niels H. Abel, a Norwegian mathematician.

Additive identity A number that, when added to any other number, does not change its value (e.g. zero).

Associative (uh-SO-see-a-tiv) pertaining to a mathematical law in which the grouping of the elements does not affect the result, e.g. 2 + (3 + 4) = (2 + 3) + 4, or 2 (3 x 4) = (2 x 3) x 4. (Compare to commutative)

Asymmetry (ay-SIM-uh-tree) Having no symmetric elements; the opposite of symmetry.

Bilateral (by-LAH-ter-ul) having, or involving, two sides; in math, arranged symmetrically on opposite sides of an axis.

Commutative (cum-YOU-tuh-tiv) pertaining to a mathematical law in which the order of the elements does not affect the result, e.g. 3 + 2 = 2 + 3, or 3 x 2 = 2 x 3. (Compare to associative)

Congruent (con-GROO-ent) (in math) when two shapes and sizes are identical.

Diameter (dy-AM-eh-tr) a straight line passing through the center of a circle, from one side to the other.

Geometric (shapes) (gee-oh-MEH-trik) refers to 2-dimensional shapes formed either by straight lines or by regular curved lines.

Group (in math) a set of elements in which any two can be combined. A group must have an identity, an inverse, closure and the associative law.

Integer (IN-tuh-jr) any positive or negative whole number or zero; not fractions.

Inverse the number that must be combined with another number to end up with the identity of a group. Example: in addition, -4 is the inverse of +4 because when you add them together the result is zero, the group's additive identity.

Mandala (MON-duh-luh, mon-DAH-luh) - a Sanskrit word meaning "circle", that has sacred significance. Mandalas may contain concentric geometric shapes, various other forms, and images of deities (religious figures). In Hinduism and Buddhism, the circle symbolizes wholeness, or the universe. Mandalas in our society are more varied and have many different meanings.

Organic (shapes) characterized by wavy lines and forms; non-geometric.

Perpendicular refers to a line that is at right angles to another line.   These 2 lines are

perpendicular to each other.

Point an element in geometry having a definite position, but no size or shape.   • A point.

Right angle an angle that measures 90 degrees, formed by two lines that are perpendicular to each other.

Right angle

Symmetry (SIM-uh-tree) repeated forms that follow one or more systems or rules; symmetry may be rotational, bilateral, translational or expansional.

Bini Mask

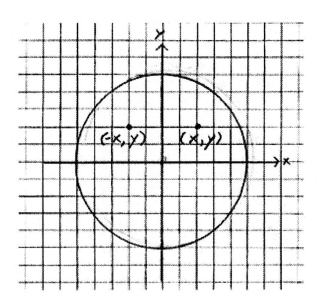

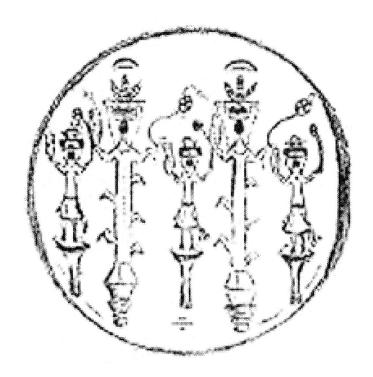

Navajo Sand Painting

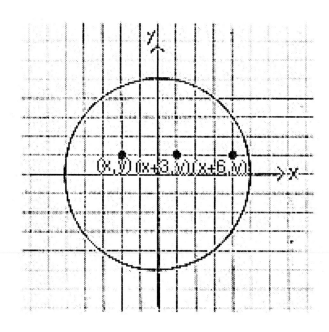

Rose Window

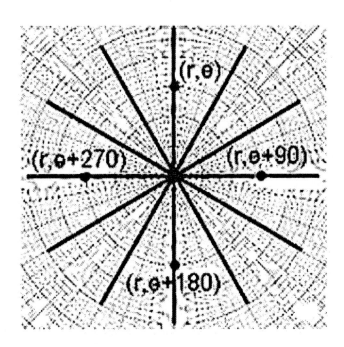

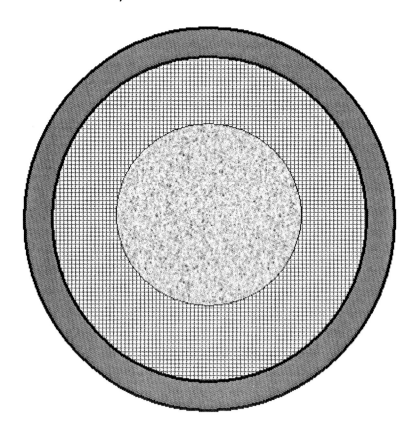

Target

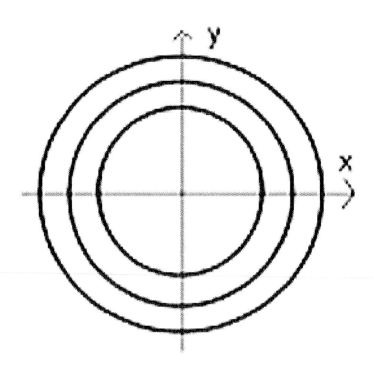

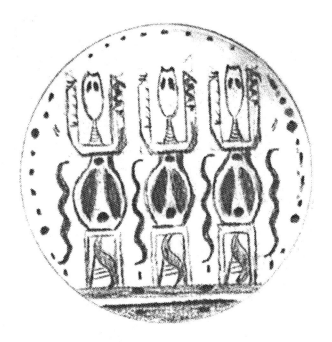

Australian Aboriginal Bark Painting

Haida Motif

Chapter 3
KITCHEN FLOORS
AND THE
ALHAMBRA PALACE

At a glance...

Time: Three 1-hour sessions, or one 1-hour session and one 2-hour session, or one 3-hour session; 1 homework assignment

Activity

M. C. Escher, Victor Vasarely and Bridget Riley are all artists who have explored *tessellations*, continuous and potentially infinite patterns of shapes, often seen in mosaic designs or tiled floors. Mathematicians since Archimedes have devoted themselves to trying to figure out what kinds of polygons could be used to "tile the plane." In this lesson, students learn about this process from the point of view of both mathematician and artist, exploring what kinds of shapes lend themselves to "tiling", and creating their own tessellated artwork. Time is also provided for a study of this technique as used by the Moors (whose religion prohibited the depiction of human figures) to create magnificent mosaics, fine examples of which can be seen at the Alhambra.

Mathematical Concepts

♦ Identification of various types of polygons

♦ Geometric vs. organic shapes

♦ Angle measurement; interior and exterior angles

♦ Symmetry

♦ Area

♦ Translation, rotation, reflection, glide reflection

Pre-planning

Read through the entire lesson ahead of time in order to know where it's going. Decide how you wish to divide the lesson or if you want to have one in-depth session. For example, you might want to think in terms of an entire morning spent on a math/art project. Look around your classroom and note any examples of tessellations on the floor, ceiling or elsewhere.

Prepare the models that will be used to demonstrate by cutting out the patterns provided and gluing them to cardboard, poster board, oak tag or other appropriate material.

Materials/Equipment

1. Overhead projector
2. Transparencies made from our masters
3. Models made from the patterns provided
4. Regular geometric shapes (meaning that all sides and angles are equal):
   - equilateral triangles, one per student
   - squares (quadrilaterals), one per student
   - pentagons, one per student
   - hexagons, one per student
   - octagons, one per student
5. Pencil with eraser (one per student)
6. Masking tape
7. Scrap paper, one sheet per student, 12"x18"
8. Crayons, colored pencils, paints or other coloring materials
9. Paper to draw and color on, 12"x18"
10. Protractors (optional), one per student

Step I – Introduction

Show transparencies of the drawings and photographs of the Alhambra Palace. The following information will help introduce Islamic art.

In the Middle Ages, Granada, Spain was the last Moorish (Arab) stronghold in Western Europe. The Alhambra was an ancient citadel, or fortress, in Granada and the residence of the Moorish monarchs. Mohammed I (1195-1272) began work on the Alhambra around 1232. Yusuf I completed the construction .project between 1333 and 1354 and his successor, Mohammed V, finished the decorations. It took about two hundred years to accomplish the entire job. The Alhambra was made of brick, wood and pise (stiff clay rammed between forms) laid out around a series of interior patios or courtyards. The outdoor walls are carved stucco; the interior walls are made of marble, ceramic tile and stucco. While the outside of the Alhambra looks like a forbidding fortress, inside it is more like a palace in which the designer was aiming for harmony of form - balancing and contrasting straight and curved lines.

Early Islamic art avoided religious symbols, possibly in deliberate contrast to Christian art of the same (Medieval) era. Instead, its design elements were based on geometric shapes, using organic forms that were highly stylized in decorative tracery that covers an entire surface with beautiful and sometimes complex patterns. The relationship of one shape to another was more important than the complete pattern, which was seen as decoration. The repeated shapes are similar to a repeated melodic or rhythmic pattern in music.

Early in 1492 (the year Columbus sailed for the New World), King Abu-Abdullah of Granada swore his allegiance to King Ferdinand and Queen Isabella of Spain and surrendered Granada to them, at which time the Spanish began vandalizing the Alhambra. Their grandson, Charles V, went even further, driving all Moslems out of Spain and tearing down large sections of the Alhambra to make way for his own palace, which he never finished building. Unfurnished and without a roof, the new section looked totally out of place. Upon seeing Charles' palace while visiting Spain, the 19th century American author Washington Irving called it "an arrogant intrusion." The last royal residents were Philip V and Elizabeta of Parma in the early 1700's. Late in that century, restoration began on the Alhambra.

Many artists and writers were influenced by the Alhambra. The Dutch artist Maurits Escher visited and sketched it in 1936. His work, more than anyone else's, contains repeated patterns stretching as if to eternity but, unlike the Moors, Escher used realistic organic forms such as reptiles, fish and birds as his "generating", or basic, shape. When Washington Irving stayed at the Alhambra in 1829 he was inspired to write *Tales of the Alhambra,* in which he imagined what might be going on outside his window in the palace.

Step 2 - Define tessellations

Tessellations are constructed from one or more shapes that repeat to form an infinite overall pattern; that is, one which could be extended forever in any direction. Have the students look at the floor, wall or ceiling of the classroom. The continuous squares or rectangles form a tessellation. The square or rectangle is called the "generating shape" of the tessellation.

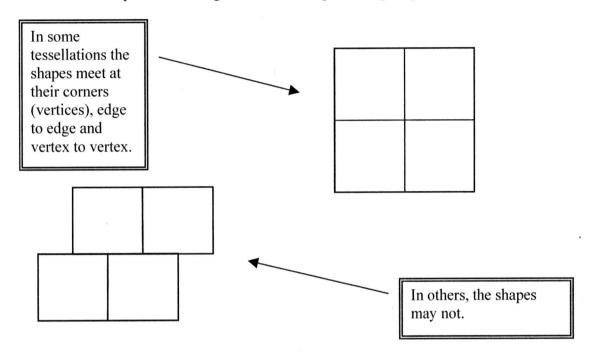

In some tessellations the shapes meet at their corners (vertices), edge to edge and vertex to vertex.

In others, the shapes may not.

The generating shape can be either geometric or organic (non-geometric).

Point out examples of each in your classroom, if possible, or using the examples provided. If the "tilings" could be <u>extended infinitely</u> in any direction, they form a tessellation.

Ask: Can you name some 2-dimensional geometric shapes?

Circle, ellipse or oval, triangle, quadrilateral, square, rectangle, parallelogram, rhombus, trapezoid, pentagon, hexagon, octagon.

Ask: Can circles tessellate by themselves?

No.

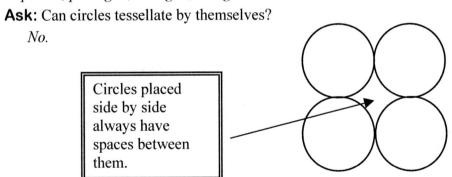

Circles placed side by side always have spaces between them.

Illustrate this on the chalkboard.

What about ovals?

No.

Illustrate as shown.

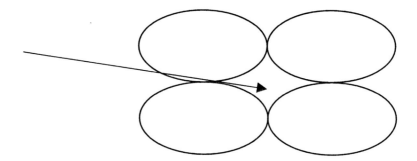

In order for a geometric shape to tessellate by itself, it must be a polygon, which is a shape that is bounded by straight lines.

Examples:

parallelogram

hexagon

Point out: All the shapes named above are polygons except the circle and the ellipse (oval). However, not *all* polygons can tessellate.

Tessellations on the Alhambra

Look at the transparencies of the Alhambra again.

With your students, note the different kinds of tessellations – those formed by shapes that tessellate by themselves and those that tessellate only with the "help" of another shape.

Court of the Fish Pond

Court of the Lions

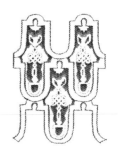

Hall of Ambassadors

Step 3 - Study polygons.

Point out: Polygons are many-sided figures. In making tessellations, some generating shapes are <u>regular</u> polygons, which means that all their sides and angles are equal.

3-sided polygons (Triangles)

Show the transparency, "The Journey From Scalene to Regular"

Ask: When is a triangle regular?

A triangle is regular when it is an equilateral triangle, which means, literally, that all sides are of equal length.

Have the students draw scalene, isosceles and equilateral triangles, noticing that the isosceles has only two equal sides, and that the scalene has no equal sides. They will find it difficult to draw a true equilateral triangle unless they use a protractor, because all three angles *must* be exactly 60 degrees. They will also see that if the sides are of equal length, the angles also must be equal.

4-sided polygons (Quadrilaterals)

Show the transparency, "Evolution of a Square."

Ask: When is a quadrilateral regular?

A regular quadrilateral is better known to us as a square, which has four equal sides and four equal angles.

Again, have your students follow, by drawing them, the two alternate routes that a miscellaneous scalene quadrilateral can take as it evolves, first into a parallelogram, then into either a rhombus or rectangle, and finally, to their common destination -- a square. Therefore, the only regular quadrilateral is a square.

> **NOTE:** While all triangles and quadrilaterals will tessellate by themselves, for the purpose of this lesson we will focus on equilateral, or regular, triangles and regular quadrilaterals, or squares.

Step 4 – Math Lesson

Distribute regular polygons to the class.

(Squares, equilateral triangles, regular pentagons, regular hexagons, regular octagons)

Give one to each student. Each student will also need scrap paper.

Using the scrap paper, the students can try to make a tessellation by repeatedly tracing around one of the shapes, edge to edge and vertex to vertex.

Have them cover a large section of the page (repeating the shape at least 10 times) to test whether the pattern is infinite, i.e. if it could continue forever. Some will work and others will not.

The students will soon discover that equilateral triangles, squares and regular hexagons tessellate by themselves...

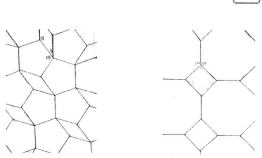

but regular pentagons and regular octagons need another shape to form a tessellation.

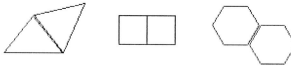

Suggested questions for review: We have grouped the questions by shape, but you may wish to jump from one shape to another spontaneously rather than focusing on them one at a time.

A. Shapes that will tessellate:

Triangles - Show a student's example of tessellated triangles to the class. (See hexagon below.)

Ask: Did they tessellate by themselves? *Yes*

Why can they do this?

Look at the angles that meet at the vertex of the tessellation. How many degrees are in each angle of an equilateral triangle? *60 degrees (Triangle has 180)*

How many 60 degree angles are meeting at the vertex? *6 angles*

What is the product of 6 (angles) x 60 (degrees)? *360 degrees*

So how many degrees are there in a circle? *360*

Squares
Go through the same process.
Look at how the squares tessellate.
Ask: How many degrees are there in each angle at the vertex where the 4 squares meet? *90 degrees*
4 (angles) x 90 (degrees) = *360 degrees*

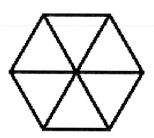

Hexagons
Ask: How can we figure out how many degrees there are in each angle of a hexagon?
Divide the hexagon into 6 equilateral triangles.
Review: How many degrees are there in each angle of an equilateral triangle? *60 degrees*
So the angle of the hexagon is 60 x 2, or 120 degrees.

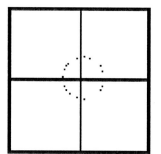

If we have 120 degrees at the vertex where the three hexagons meet, 120 x 3 = *360 degrees.*

B. Shapes that will not tessellate:

Regular Pentagons

Using a student example or our master, point out the leftover (negative) spaces between the shapes.

Ask: Why doesn't a regular pentagon tessellate by itself?

How come the other shapes tessellate by themselves and this one does not?

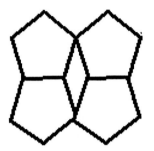

Because the number of degrees in the <u>angles at the vertex where the shapes meet</u> do not divide evenly into 360.
(Each angle is 108 degrees)

Ask: What is the shape that is left over between the pentagons? *A parallelogram.*

Regular Octagons

Go through the same process, demonstrating that the space between the octagons is a square. (Each angle of the octagon is 135 degrees. 2(135)=270. 360-270=90, which is the number of degrees in each angle of a square).

BREAK THE LESSON HERE

KITCHEN FLOORS continued...

Step 5 – Art Activity

Show the 3 overhead drawings of Mandelbaum's designs.

Point out: All are based on polygons that tessellate by themselves: equilateral triangles, squares, and regular hexagons. BUT If we modify any of these shapes in certain ways, maintaining the same area, <u>the new shape, even if it is no longer a geometric shape, will also tessellate</u>.

Students create organic shapes that will tessellate.

Distribute the following: squares or hexagons (1 per student); poster board; pencils with erasers; scissors; tape.

    a. Have the students trace around the squares or hexagons and cut them out.

    b. The students then place the cut-out shape on their desks and draw any simple curve (no overlaps) that connects the vertex (corner) on the upper left to the vertex on the upper right. *Use our <u>models</u> to demonstrate <u>or draw the diagrams</u> on the board.

RIGHT WAY   WRONG WAY

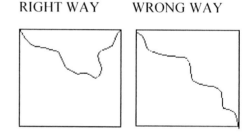

Note: Make certain that your students draw the curve as illustrated as they sometimes make a mistake and skip over a vertex.

    c. The shape is then carefully cut out in one piece, along the curve. Show the class how the piece fits into the square; then "glide" to the bottom and tape it there. This is called a translation, because you are "translating" the shape to the opposite side.

Note: This process works because the opposite sides of a square are equal. The opposite sides of a regular hexagon are also equal.

d. Now a simple curve is drawn from the upper left vertex to the lower left vertex, or corner, as shown.

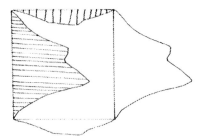

e. The shape is again cut out, glided or translated to the opposite side, and taped as before. The result is an organic shape which is equal in area to the original geometric shape. This shape will tessellate.

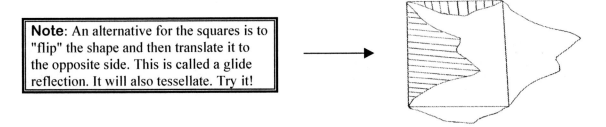

Note: An alternative for the squares is to "flip" the shape and then translate it to the opposite side. This is called a glide reflection. It will also tessellate. Try it!

If you are using a hexagon, repeat steps b through e with the third pair of opposite sides.

Students are now ready to create original tessellation designs.

a. Distribute the 12" x 18" drawing paper
b. Students draw their own tessellations by tracing around the organic shapes they have created, continuing the design to the edges of the paper. They first trace around their organic shape one time any place on the paper. Then they translate the shape to the right or left, up or down, and find that the pieces will fit together like a puzzle.
c. The tessellations are colored.

BREAK THE LESSON HERE

KITCHEN FLOORS continued...

Step 6 – Additional Art Activity: A more complicated way to make tessellations

Distribute squares or hexagons, one per student.

Demonstrating with the model provided, draw a simple curve from the upper left to the upper right vertex of the hexagon. Have your students draw a similar curve. (It need not be identical to yours.)

Students cut out the curve in one piece.

They lay the hexagon flat on the desk and fit the cutout shape back in.

"Swing out" the cut-out shape and rotate it around the vertex to the adjacent side. This must be done without lifting the shapes off the desk. Be certain that the students have not reflected the shape (flipped it over) before rotating it. Demonstrating using the model will help avoid confusion.

Repeat this process with the next two sides as illustrated, and again with the last two. The result is an organic shape with the same area as the original geometric shape. This shape will *still* tessellate.

Students create original tessellation designs.

  a. Distribute the 12" x 18" drawing paper

  b. Students draw their own tessellations by tracing around the organic shapes they have created, continuing the design to the edges of the paper.

  c. The tessellations are colored.

Compare the first group of tessellations with the second.

  ♦ Since the generating shape of the first group was made by translating the cut-out section to the <u>opposite</u> side, the tessellation is formed by translating the shape to opposite sides.

  ♦ Since the generating shape of the second group was made by rotating the cut-out section around the vertex to the <u>adjacent</u> side, the tessellation is formed by rotating the shape around a vertex.

  ♦ The first one appears to "move" horizontally, vertically or diagonally; the second appears to move in a circle.

Homework assignment: Perhaps a floor, ceiling or wall in your home has tessellations. Maybe an outside wall or roof has tessellations. Can you think of one? What is the generating shape or shapes? Do the shapes meet edge to edge and vertex to vertex? Is the generating shape geometric or organic? Bring in a drawing of the tessellation in your home to show the class.

Note: If you wish to extend the lesson, you can ask the students which 3-dimensional shapes will tessellate. *Rectangular solids.* They can also investigate, and possibly attempt to draw, what shapes are left over (negative shapes) when you attempt to tessellate using other solids.

Tangents: Related Activities for the Core Curriculum
Language Arts
1. Read Washington Irving's *Tales of the Alhambra*. Students can discuss and write about what they might imagine outside their own windows at night, either in the present or in centuries past.
2. Read and discuss additional works by Irving. What geographic area did he write about most often? (Towns along the eastern side of the Hudson River.)
3. Look up all the meanings and uses of the word *square*. Use the word square in every sense of the word: "un-hip," town square, "square peg in a round hole" etc. Do these various meanings relate to each other in any way? Make up a story about any type of square.
4. Look up the different meanings of *reflection* and *translation*. Write a story using one of the meanings.

Social Studies
1. Research Arab cultures in Spain and elsewhere. How is the influence of Moorish culture still seen in modern Spain? Are there any local examples?
2. Discuss any buildings or neighborhoods in the vicinity of the school that were torn down and rebuilt or put to new uses. Are there existing unused structures (e.g. an old movie theater, school or warehouse) that could be preserved and used for a different purpose?

Science
Look up the kinds of designs made by sound waves. Sound waves can stretch to infinity, but do they tessellate?

Music
Listen to the music of Steve Reich or Philip Glass. Note the endlessly repeated melodic patterns. Do they seem to relate in any way to visual tessellations?

Visual Art
Research the work of Escher, Riley and Vasarely. Discuss how their tessellations were similar to or different from Islamic tessellations and one another's.

Stefanie Mandelbaum and Jacqueline S. Guttman

For Further Reference

Jill Britton, *Introduction to Tessellations*, Dale-Seymour Publications, 1989

Keith Critchlow, *Islamic Patterns*, Schocken Books, 1976

Issam El-Said and Aysa Parman, *Geometric Concepts in Islamic Art*, World of Islam Festival Publishing Co., Ltd., 1976

Bruno Ernst, *The Magic Mirror of M.C. Escher*, Ballantine Books, 1976

Doris Schottschneider and Wallace Walker, *M.C.Escher Kaleidoscopes*, Tarquin Publications, 1978

There is a computer program called Tesselmania, for which this lesson is a good introduction.

Glossary

Adjacent adjoining, near.

Allegiance the loyalty of a citizen to his government or of a subject to his sovereign.

Generating shape the shape which is repeated, either by rotation or reflection or translation, in order to form a tessellation.

Infinite the boundless regions of space; unbounded or unlimited; immeasurably great; indefinitely or exceedingly great.

Moorish of or pertaining to the Moors (a Muslim of the mixed Berber and Arab people inhabiting NW Africa; a member of this group that invaded and conquered Spain in the 8th century).

Opposite face to face.

Organic refers to irregular shapes which suggest forms found in nature.

Pise (peez) a mixture of sand, loam, clay, and other ingredients rammed hard within forms as a building material. Also called rammed earth.

Reflection mirror image.

Rotate to turn around a given point.

Stylized designed to conform to a particular style or kind of art, for example, baroque style or abstract style.

Tessellation a design formed by the potentially infinite repetition of a given shape (see generating shape).

Tracery a delicate, interlacing design.

Translation the movement of a given shape in a straight line, either diagonally, horizontally or vertically.

Vertex the intersection of two sides of a plane figure (the corner point).

Court of the Lions, Alhambra

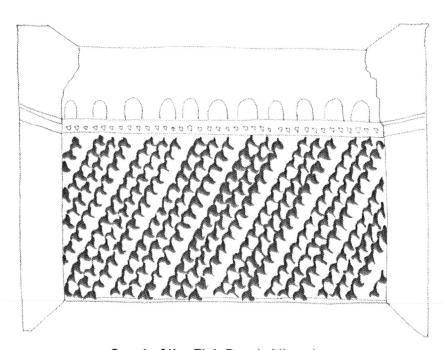

Court of the Fish Pond, Alhambra

Ceiling, Court of the Fish Pond, Alhambra

Tessellation in the Baths, Alhambra

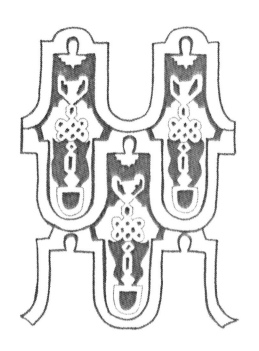

**Hall of Ambassadors,
Alhambra**

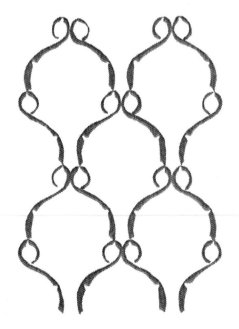

**Court of the Mosque,
Alhambra**

Regular Pentagon Diagram

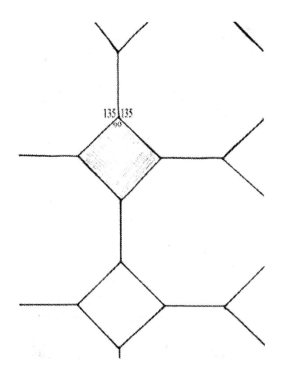

Regular Octagon Diagram

Stefanie Mandelbaum and Jacqueline S. Guttman

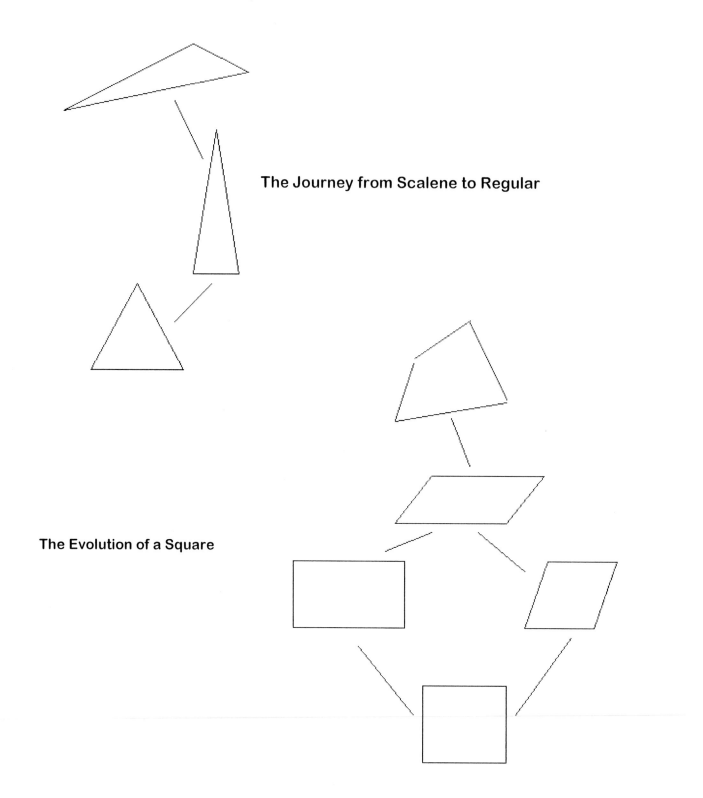

The Journey from Scalene to Regular

The Evolution of a Square

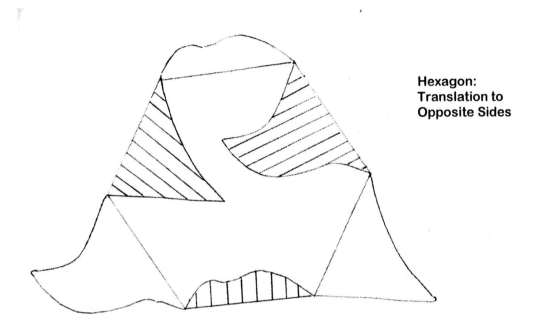

**Hexagon:
Translation to
Opposite Sides**

**Hexagon:
Rotation about
Vertices**

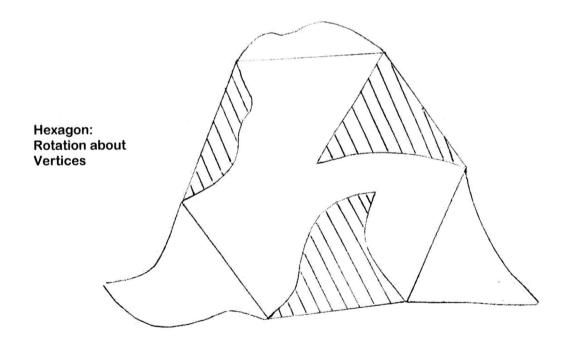

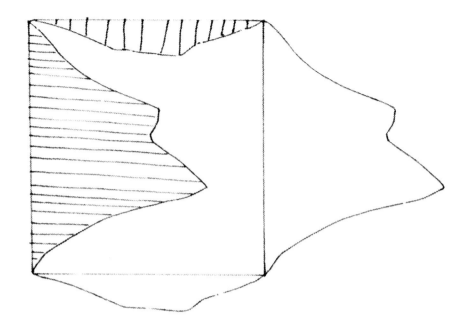

Square: Translation to Opposite Sides

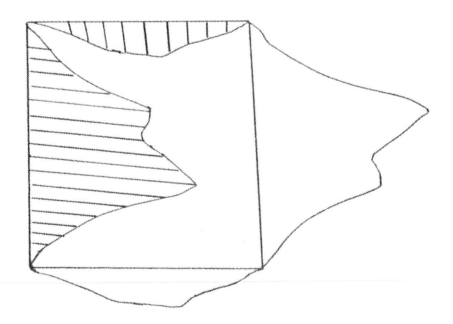

Square: Translation and Glide Reflection to Opposite Sides

Chapter 4
COLLAGES AND COMPOSITES

At a glance...

Time: One 90-minute session followed by several short drills.

Activity

Students construct collages whose elements are actually factors of composite numbers. In the course of the lesson they learn about or review the differences between prime and composite numbers. They also are introduced to artists who are known for their work with collages.

Mathematical Concepts

♦ Working with fractions, decimals and percentages; converting them to and from each other.

♦ Finding the factors of composite numbers.

Pre-planning

Read through the entire lesson ahead of time in order to know where it's going. Read through the glossary. Decide whether you prefer to divide the lesson into two sessions (art project alone) or complete it in a single session (art project followed by math drills). You might want to think in terms of a morning spent on math and art.

Masters/transparencies should be duplicated in advance.

Materials/Equipment

1. Glue sticks, one per student
2. Scissors, one pair per student
3. 9"x12" black construction paper, one sheet per student
4. Copies of our pattern master on bright colored copy paper (six different colors), one per student (For easier cutting, you may wish to enlarge the pattern.)
5. Pencils
6. Notebook paper
7. Transparencies made from our masters
8. Overhead projector

Step 1 - Art Introduction

Begin by showing the transparencies of Stefanie Mandelbaum's "neo-cubist" and organic collages.

The following information should be helpful:

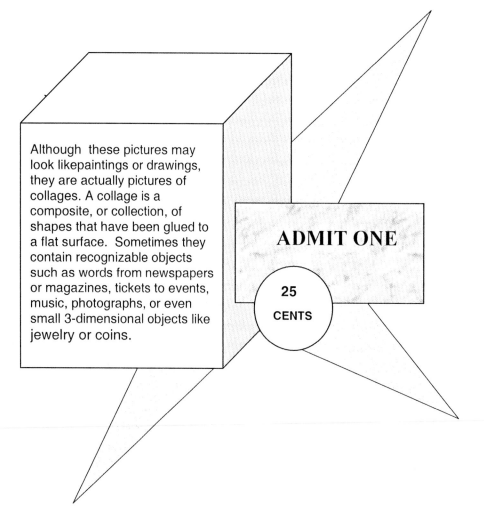

Although these pictures may look likepaintings or drawings, they are actually pictures of collages. A collage is a composite, or collection, of shapes that have been glued to a flat surface. Sometimes they contain recognizable objects such as words from newspapers or magazines, tickets to events, music, photographs, or even small 3-dimensional objects like jewelry or coins.

ADMIT ONE

25 CENTS

Some artists who constructed collages, like Pablo Picasso (1881-1973), Georges Braque (brahk) (1882-1963) and Juan Gris (gree) (1887-1927), based much of their art on the shapes created from a perspective drawing of a cube or other geometric solid. This was called Cubism (CUE-bism). Cubism is a style of painting and sculpture developed in the early 20th century which reduced natural forms to their geometrical equivalents (cylinders, cubes, prisms, etc.) It was greatly influenced by African art. Synthetic Cubism was a collage form of cubism begun in 1912. The collages were made from various geometric shapes arranged to represent parts of a subject. A Picasso collage called *Still Life with Chair-Caning*, made of oils, rope and pasted paper, was the first in this style. Braque followed with *Fruit Dish and Cards*. Picasso and Braque worked very closely together. Sometimes their work was so similar that even the artists could not remember who had done what! The Cubists also used the idea that each object could be seen from a different point of view or various points of view simultaneously.

Step 2 - Math Introduction

Review composite numbers.

Ask: What is the difference between prime and composite numbers?

A prime number is divisible only by itself and 1.

Ask: Who can name all the composite numbers from 1 to 20?

4, 6, 8, 9, 10, 12, 14, 15, 16, 18, 20

Ask: What are the factors of each of these composite numbers? (Write the answers on the board).

4 = 1x4, 2x2, 4x1

6 = 1x6, 6x1, 2x3, 3x2

8 = 1x8, 8x1, 2x4, 4x2

9 = 9x1, 1x9, 3x3

10 = 1x10, 10x1, 2x5, 5x2

12 = 1x12, 12x1, 2x6, 6x2, 3x4, 4x3

14 = 14x1, 1x14, 2x7, 7x2

15 = 15x1, 1x15, 3x5, 5x3

16 = 1x16, 16x1, 2x8, 8x2, 4x4

18 = 1x18, 18x1, 2x9, 9x2, 3x6, 6x3

20 = 1x20, 20x1, 2x10, 10x2, 5x4, 4x5

Step 3 - Art-Math Activity

- ◆ Seat students in groups of six. Distribute the art materials. Each student should have a pair of scissors, a glue stick, one sheet of black construction paper, and a colored sheet of numbered rectangles copied from our master.

- ◆ Cut out the numbered rectangles and then place them in separate piles in the center of the table: one pile for all the 2s (of any color), one pile for all the 3s, etc.

- ◆ Assign a composite number to each student from the numbers on the board.

- ◆ Each student collects from the center of the table <u>all</u> of the rectangles whose numbers are factors of his/her assigned composite number. For example, if a student was assigned the number 10, the student collects one 10, two 5s, five 2s, and ten 1s.

♦ Students glue each group of factors on the black construction paper, using whatever colors, positions and locations satisfy their sense of balance and beauty. Some examples:

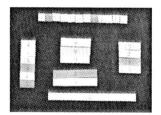 

The collages can now be used in various ways to discuss fractions, decimals and percents, as will be shown.

BREAK THE LESSON HERE

COLLAGES AND COMPOSITES continued...

Step 4 - Discussion

Select a student's work as an example.

Note: Since the students will have chosen different color combinations, each collage will have different answers to each of these questions. We have used circles instead of rectangles, to demonstrate that this lesson may be taught using any shape.

Examples: Suppose you have decided to discuss collages that are based on the composite number 18. (6x3, 3x6, 9x2, 2x9, 18x1, 1x18)

Suggested questions for Example A, or 6x3:
What fraction of the shapes in the rectangle is <u>black</u>? *1/3*
What decimal does 1/3 equal? *0.333....*

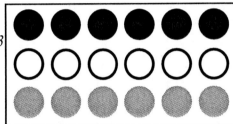

Look at Example B, or 9x2:
What fraction is <u>gray</u>?   *1/3*
What fraction is <u>black</u>?   *4/9*
What fraction is <u>white</u>?   *2/9*
Add: 1/3 + 4/9 + 2/9 = *9/9, or 1*

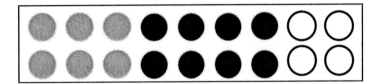

Look at Example C, or 2x9:
What percentage of the rectangle is <u>white</u>?
9/18 = 1/2 = *50%*

1x18

Note: The same mathematical concepts can be taught with more artistic flexibility. Some suggestions:

Pattern Blocks

These or other objects (e.g. coins) can be traced around onto oak tag or other heavy paper and cut out. Using these patterns, the shapes can then be traced onto different colored construction paper. Each student works with only one factor combination, e.g. 3 x 6 =18. S/he cuts out 18 of the same shape in 2 or more colors and arranges them three down and six across. The same questions can be asked: What percent of the shapes are orange? and so on. (Another student at the same table may use 9 x 2 =18, and a third student 1x18=18.)

Organic Shapes

Henri Matisse (1869-1954) created collages by first painting paper to get the colors that he wanted and then cutting out and pasting various shapes onto a background. His soft, wavy, organically-shaped collages are loved by many people. Students can use his technique or draw an original organic shape onto heavy paper and cut it out to use as a pattern which is, in turn, traced onto colored construction paper.

Once the concept of factoring composite numbers is grasped, the construction of collages can be taken even further. Some suggestions:

Self-portraits and Family Composites

By tracing their patterns onto family photographs, students can cut them out and create personal collages.

Magazine Collages

Rather than using family photographs, the patterns can be traced onto pages torn from magazines.

Collage with a Geometric "Frame"

Pattern blocks or other templates can be used to trace around on a sheet of construction paper that becomes a "mat" (see illustration). After the collage is complete, the mat is placed over it to make a "frame."

  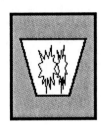

Tangents: Related Activities for the Core Curriculum

Language Arts

1. If students create "personal" collages about themselves and their families, they can then create oral histories or essays that are related to the images in the collage.
2. Have the students look up the many meanings and contexts of the words "composite", "prime" and "factor." How many ways can they use them in a sentence? What are their derivations and roots? What other words are they related to? (composing, composition, compost, compound; primary, premier, primogeniture, primer – as a first reader or an underlayer of paint; fact, faction, factory, face, facet, etc.) Can they use these different words in a sentence or paragraph?
3. Go through the same exercise with the prefix "neo-", meaning new or recent, in order to have a better understanding of "neo-cubism."

Social Studies

1. Using a recent newspaper or magazine issue, students can create a current events collage. This can lead to a class discussion of the composite of articles and photos that are in the collage – and to learning about a newspaper "composing room."
2. The Cubists were influenced greatly by African sculpture. In 1920, Juan Gris published an essay on this subject pointing out how the distorted proportions found in much African work influenced western art such as Picasso's *Demoiselles d'Avignon*. Students can research examples of African sculpture from different nations, noting its symbolism and place in daily life.

Science

Look at cutaways of, for example, the earth's crust, a chocolate layer cake, a "Dagwood" sandwich, etc. Stacking different things atop one another and cutting away a section will leave students with -- Presto! -- a collage.

Design

1. The clothes we wear are a composite of things we have selected from our closets and drawers. Why do we make certain clothing choices? Students can bring in pieces of fabric and decide which ones work well together and why.

2. Students can assemble a rock garden, which is a composite of rocks, plants and flowers. If space is not available, this can be done in miniature, with stones, leaves and twigs.

Culinary Arts

Arrange food or snacks on a plate by shape, color or texture so that they form a collage that is pleasing to the eye.

For Further Reference

Maly and Dietfried Gerhardus, *Cubism and Futurism*, Phaidon Press Limited. Oxford, 1979

Jean Selz, *Matisse*, Crown Publishers, Inc., New York (no date of publication)

John Elderfield, *The Cutouts of Henri Matisse*, George Braziller, New York, 1978.

The Nelson A. Rockefeller Collection, *Masterpieces of Primitive Art*, Alfred A. Knopf, New York, 1978.

Glossary

Collage (co-LAZH)   a technique of composing a work of art by pasting on a surface various materials not normally associated with one another, such as newspaper clippings, theater tickets, fragments of an envelope, etc.

Composite (com-PAH-zit) made up of disparate or separate elements; compound.

Cubism, Cubist a style of painting and sculpture developed in the early 20th century which reduced natural forms to their geometrical equivalents, (cylinders, cubes, prisms, etc.)

Factor (in math)   a divisor; one of two or more numbers which, when multiplied together, produce a given product.

Henri Matisse (1869-1954) did paintings made of gouache (opaque watercolor) on cut and pasted paper, a form of collage that used organic shapes, instead of geometric ones.

Neo-Cubist a word coined by S. Mandelbaum to describe some of her collages, which are different from cubism, but derive from the same idea. (Picasso's and Braque's collages were called cubist by a critic who didn't like them and was being sarcastic.)

Prime number not divisible without remainder by any number except itself and one.

Synthetic cubism (art) the late phase of cubism characterized chiefly by an increased use of color and the imitation or introduction of a wide range of textures.

Neo-Cubist Collage

Neo-Cubist Collage 2

Duet

Organic Collage

20	18	16	15	14	12	10	9	8	7	1
20	18	16	15	14	12	10	9	8	7	1
										1
										1
										1
										2
										6
									1	
								4	1	
							3		1	
						2			1	
									1	
					5	4	5	2	1	1
									1	2
								4	1	
				6					1	2
			3						1	
		1				2			1	
		1			3		3		1	2
	2	1	2			2		2	1	
		1							1	1

69

Chapter 5
THE GOLDEN MEAN

At a glance...

Time: Two 1-hour sessions or one 2-hour session

Activity

Many artists and architects have used the Golden Mean and/or Fibonacci numbers as the mathematical basis for their work. In this lesson, students will learn about golden rectangles, triangles and ellipses and see examples of some of the ways in which the Golden Mean was used by the Romans. Following that, they will discover the pattern inherent in the Fibonacci sequence, the relationships between Fibonacci numbers and the Golden Mean, and examples of how both are found in nature. Finally, they will create collages and rubbings derived from various golden geometric shapes.

Mathematical Concepts

- ♦ Golden rectangles, triangles, and ellipses; isosceles triangles
- ♦ Proportions
- ♦ Addition, subtraction, multiplication, division, fractions, decimals
- ♦ Use of calculators
- ♦ Number sequences: finding patterns in sequences of numbers
- ♦ Measurements

Pre-planning

Read through the entire lesson ahead of time. Decide whether you prefer to divide it into two sessions or complete it in a single, 2-hour session. Assemble the necessary materials. If necessary, modify the lesson according to the abilities of your class. Transparencies and photocopies should be prepared in advance.

Materials/Equipment

1. Transparencies made from our masters
2. Copies of our masters of shapes that have "golden" proportions (packet of 3 sheets), one packet per student
3. 9"x12" drawing paper, a few sheets per student
4. Pencils with erasers

5. Colored chalk, one piece per student
6. Scissors for each student, if possible
7. Glue sticks, one per student
8. Calculators
9. Overhead projector

Step 1 – Introduction

The following information should help you describe the golden mean.

The proportions that we call the Golden Mean have been considered ideal, or perfect, for thousands of years. Examples of the Golden Mean can be found throughout the world, in many different cultures and eras. The pyramids of ancient Egypt and much Greek and Roman architecture and sculpture were based on these proportions. During the Renaissance (14th and 15th centuries) Leonardo da Vinci (lay-oh-NAR-do da VIN-chee) illustrated a book about the Golden Mean, and many European artists, including Veronese (vay-roe-NAY-say), Rembrandt, and Seurat (suh-RAH), based their work on it. In the 20th century, the artist Mondrian was also captivated by these magical proportions.

Show transparencies illustrating some of the ways the golden mean has been used by artists and architects. Describe the Colosseum:

The Greeks believed in moderation in all things, a philosophy that is reflected in their use of "golden" proportions, which do not allow for either long narrow shapes or short wide ones. However, after the Romans conquered Greece, they, too, adopted the Golden Mean, but did not follow the idea of moderation in other parts of their lives. One of the best examples of the Roman use of the Golden Mean was in the Flavian Amphitheatre, which we know better as the Colosseum. Begun in the year 72 and completed 8 years later, it is a triumph of engineering and architectural skills. The shape is elliptical, or oval, and measures an enormous 205 x 170 yards. (By comparison, a football field is 100 yards long.) The Colosseum seated 45,000 people. Around the outside there were three rows of columned arches, all of which have 'golden' proportions.

Its use, however, was far from "golden." The Colosseum was the scene of many terrible events -- violent so-called "games" that involved the killing of both animals and people. The floor of the arena, usually covered with sand, was sometimes even filled with water and used for imitation naval battles. Underneath this were prisons for animals and people. As this was during the last years of the Roman Empire, the idea was to develop a war-like spirit among the people. Eventually, the violence became so extreme that one emperor tried to put a stop to these activities, but the Romans were so accustomed to being entertained in this way that they did not want to give it up. At the beginning of the 5th century, a monk who entered the arena to try to make the games end was actually stoned to death, but from that day on, the dreadful shows stopped.

Much of the Colosseum still stands, a tribute to the skills of an ancient society. Even today, it is the pride of Rome.

Step 2 – Math and the Golden Mean

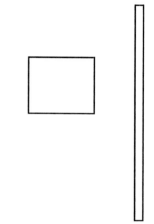

Using a rectangle, demonstrate the mathematics involved. Explain that rectangles can be of different proportions, ranging from a square with all sides of equal length to a very narrow rectangle whose length is almost infinitely long and whose width is almost zero. Draw examples on the chalkboard as shown.

A golden rectangle is visually somewhere between these two examples. Its formula is: width is to length as length is to width *plus* length. W:L = L:(W+L) Point out that, interestingly, these proportions are often found in nature. For example, an egg fits within a golden rectangle. Working out the actual mathematical formula is very complicated, but the class can be given the information that if the width of a rectangle equals 1 inch, then the length of that rectangle, if it is *golden*, is approximately 1.62 inches.

Distribute calculators to the class in order to work with these proportions.

Ask: How long would the length be if the width were 2 inches?

3.24 (2x1.62=3.24)

How long if the width were 3 inches?

4.86 (3x1.62=4.86)

Step 3 - Introduce Fibonacci

During the Middle Ages, a mathematician named Fibonacci discovered a sequence, or pattern, of numbers which relates both to nature and the Golden Mean. Who was this man?

Leonardo Fibonacci (lay-oh-NAR-do fee-boh-NAH-chee), (also Leonardo of Pisa) was born in Pisa, Italy around 1175. His name means "Leonardo, son of Bonaccio (boh-NAH-cho)," just as Leonardo da Vinci means Leonardo from Vinci. His father's business caused him to travel on long trips to Egypt, Sicily, Greece and Syria. He often took Leonardo with him, where the boy, who was very good at mathematical calculations, learned a lot about Eastern and Arabic ways of solving math problems. By the time he was in his twenties, Fibonacci was certain that the Hindu-Arabic methods of calculating were better than what he had learned at home. In 1202 he published a book about arithmetic and elementary algebra called the *Liber Abaci (LEE-br ah-BAH-chee),* or Computation Book. The book became very famous and helped introduce the numbers we use today (Arabic numerals) into Europe.

Fibonacci's examples of problems and their solutions influenced other mathematicians for hundreds of years. Once Fibonacci was invited to participate in a math tournament at the court of Emperor Frederick II. Everybody was amazed at his great talent because very few people at that time had such a remarkable understanding of mathematics. One of the problems that appeared in the *Liber Abaci* we still sometimes see in books today. He wrote it like this: "There were 7 old women on the road to Rome. Each woman has 7 mules; each mule carries 7 sacks; each sack contains 7 loaves..." Today, we are more familiar with an Old English children's puzzle that has a trick at the end:

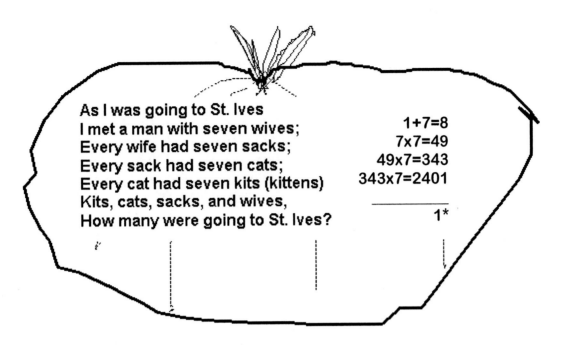

As I was going to St. Ives
I met a man with seven wives;
Every wife had seven sacks;
Every sack had seven cats;
Every cat had seven kits (kittens)
Kits, cats, sacks, and wives,
How many were going to St. Ives?

1+7=8
7x7=49
49x7=343
343x7=2401
_____
1*

*Why only 1? Look again! *Who* was going to St. Ives?

Ask: How many were going in the other direction? (The man, wives, etc.)

2,801

Fibonacci Numbers

Fibonacci discovered a sequence of numbers that, strangely enough, appears in many different ways in both math and nature. The sequence is 0, 1, 1, 2, 3, 5, 8, 13... It grew out of this problem:

How many pairs of rabbits descended from just one original pair can be born in a year if:

  a. each month each pair gives birth to a new pair and

  b. in the second month each new pair becomes mature and able to reproduce?

Include the Original Pair in your calculations.

○ = immature	● = mature	TOTAL NO. OF PAIRS
	Original Pair	1 pair
JAN.	same pair, ready to reproduce. No new pairs yet.	1 pair
FEB.	Original Pair 1st descended pair from the original	2 pairs
MAR.	Original Pair 1st descended pair - mature 2nd descended pair from original	3 pairs
APR.	Original Pair 1st descended pair 2nd desc. pair from original 1st desc. pair from 1st desc. pair 3rd desc. pair from orig.	5 pairs
MAY	Orig. Pair 1st desc. pair fr. orig. 2nd desc. pair from orig. 1st desc. pair from 1st desc. pr. 3rd desc. pair from orig. 1st desc. pair from 2nd desc. pair fr. orig. 2nd. desc. pair from 1st desc. pr. 4th desc. pair from orig.	8 pairs

Ask: Can they discover the pattern in the sequence? Can anyone figure out what the next number in the sequence would be?

Adding any two adjacent numbers in the sequence gives you the number that should come next. For example: 3+5 = 8; 5 + 8=13; 8+13 = 21; 13+21 = 34. Coincidentally, many forms in nature are examples of Fibonacci numbers, such as the number of petals on daisies (21 or 13, depending on the variety), iris (3), buttercups (5) or the number of legs on a starfish (5).

Fibonacci and Division

Have the students use their calculators to do some division examples that are based on Fibonacci's number sequence and write their answers on the chalkboard. Examples: 2/1 (=2), 3/2 (=1.5), 5/3 (=1.67), 8/5 (=1.60), etc.

Ask: What numbers should we divide next?

13/8 (=1.63), 21/13 (=1.615), 34/21 (=1.629), etc.

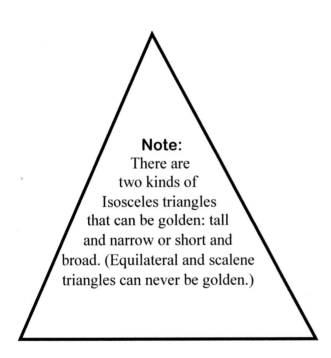

Note:
There are
two kinds of
Isosceles triangles
that can be golden: tall
and narrow or short and
broad. (Equilateral and scalene
triangles can never be golden.)

Point out that the farther out you go in the Fibonacci sequence, the closer the answer will be to 1.62. That number is the quotient of the length and the width of a golden rectangle. This means that if we take two consecutive Fibonacci numbers, e.g. 8 and 5, and make a rectangle that is 8" x 5", that rectangle is approximately "golden."

Ask: Is there anything in the class that you think has the dimensions of a golden rectangle? (Try index cards, notebook paper, books, desks, etc.)

Have the students calculate the answer by 1.) measuring the length and width and, 2.) dividing the longer side by the shorter side. If the quotient is greater than 1.62, the rectangle will be too narrow to be golden. If the quotient is less than 1.62, the rectangle will be too squarish.

Step 4 - Begin working with golden shapes.

Distribute rulers and copies of the shapes made from the masters to each student. The triangles, rectangles, ellipses (ovals) and other shapes all fit into golden rectangles or have golden proportions.

Point out: All the shapes they have been given are "golden."

a. Have the students measure the proportions of one of the rectangles. Write these proportions on the chalkboard. Have the students divide the length (longer side) by the width (shorter side) to see if it is actually a golden rectangle.

b. Have them measure the long side of one of the triangles and divide it by the short side to see if it is a golden triangle.

c. Have them measure the length of the one of the ellipses and divide it by its width to see if it is a golden ellipse.

In each case the answer should be approximately 1.62. That proportion is what makes the shape golden. For some reason, humans seem to be most comfortable with golden shapes -- in windows, rooms, tables, stationery -- throughout everyday life.

BREAK THE LESSON HERE

GOLDEN MEAN continued...

Step 5 – Art Activity

Distribute shape packets, scissors and a blank sheet of paper to each student. (White copy paper is fine.) Students cut out the golden shapes, including the large golden rectangle, and glue them in any arrangement onto the sheet of paper. (It will not yet look like a piece of art. Do not be concerned about this.) If they wish to glue the smaller shapes onto the large rectangle, it will "frame" their work when they proceed to the next step.

Step 6 - Make a rubbing from the collage.

Distribute a sheet of 9"x12" drawing paper and one or more pieces of colored chalk to each student. Show the class how to center their collages under the larger sheet and rub the chalk over the whole surface very thoroughly to get a rubbing of the collage.

Step 7 - Experiment

Use various colors for each version, sometimes rubbing over all of the shapes and sometimes over just a few. Students can try different versions until they achieve the effect they desire.

Point out: When different colors are used for the same rubbing the feelings evoked will be different also.

Step 8 - Display the work

Each rubbing (or the different versions) can be displayed next to the collage from which it was generated.

Homework assignment: Worksheet to be copied from our master:

Find the golden shapes. Some rectangles and triangles and ellipses on the page are not golden. Which are? Circle them.

Tangents: Related Activities for the Core Curriculum

Language Arts

    a. Explore the uses of the word "golden," e.g. golden boy, golden age, golden opportunity. Why golden? Why not silver, or diamond? Why *golden* mean, rather than *perfect* mean?

    b. Look up various definitions of Golden Mean, which has been used in non-mathematical ways. Horace, for example, defined it as "the safe, prudent way between extremes." In the same way, the dimensions of a golden rectangle are neither extremely long nor extremely wide.

    c. Have the students also look up the several definitions of "mean." How many ways can they use it in a single story? Do the same with "extreme."

    d. Write a group story (each person adds on a sentence) about the golden mean, as it represents perfection in nature, architecture and, possibly, life.

    e. After viewing the Colosseum, Lord Byron was inspired to write the following poem, which you may wish to read and discuss with your class:

Arches on arches! As it were that Rome,
Collecting the chief trophies of her line,
Would build up all her triumphs in one dome.
Her Coliseum stands, at moonbeams shine
As 't were its natural torches, for divine
Should be the light which streams here, to illume
This long explored but still exhaustless mine
Of contemplation, and the lesser gloom
Of an Italian night, where the deep skies assume
Hues which have words, and speak to ye of heaven,
Floats o'er this vast wondrous monument
And shadows forth its glory. There is given
Unto the things of earth, which time hath bent,
A spirit's feeling, and where he hath leant
His hand, but broke his scythe, there is a power
For which the palace of the present hour
Must yield its pomp, and what till ages are its dower.

One possible exercise: "Translate" into contemporary English usage, or rap, if that is preferred.

    f. Investigate the meaning of *Liber Abaci*. What words do we know that have the root *Liber?* (library, liberal, liberate, liberty [and hence Liberia, founded by freed slaves]) Does your community have a "Free Library?" Look up the many definitions of these and related words.

Social Studies

1. Study trading in the 12th century. Where did the Europeans go to trade? What items did they bring to trade? For what other items? What was available in the Middle East that they could not get at home? What means of transportation did they use? What were their routes?

2. Learn about Marco Polo, a very famous traveler who lived in the late 1200's and early 1300's. He introduced Europeans to products from Eastern Asia. Where did he go? What did he bring back?

3. In what other ways besides mathematics were Europeans influenced by these far-off lands?

Advanced Mathematics (For gifted and talented students.)

The following is a slightly updated version of a problem that was in Fibonacci's *LiberAbaci.*

    If B gives A 7 pennies, then A has five times as many as B.

    If A gives B 5 pennies, then B has five times as many as A.

    How many pennies does each one have?

    A has 8 pennies; B has 10 pennies (This problem can be solved using simultaneous equations.)

Visual Art and Architecture

1. Look at a photograph of the United Nations headquarters in New York City. If possible, measure the dimensions. Your students will see that they are golden. Is it a coincidence that this building which houses an organization devoted to peace and harmony was designed with golden proportions?

2. Have your students research the life and work of Leonardo da Vinci, Veronese, Rembrandt, Seurat and Mondrian. How did they incorporate the golden mean into their work?

Science

1. We have noted that the petals on many flowers are one of the Fibonacci numbers. Here is a more complete list:

   a. Lily, Iris: 3
   b. Columbine, Larkspur, Buttercup: 5
   c. Delphiniums: 8
   d. Corn marigolds: 13
   e. Aster: 21
   f. Daisy: 34, 55, 84

   Have the students count the petals on any flowers in the classroom or at home. On how many of them do they find Fibonacci numbers?

2. Leaves are also often arranged on stems using numbers found in the Fibonacci sequence. For example: call the leaf nearest the base of a stem "B" (for Base). Count the subsequent number of leaves until you reach a leaf that is directly above the first one. The total number of leaves counted is usually a Fibonacci number.

Ask: Why are the leaves not arranged one under the other? *Because sunlight must be able to reach all the leaves in order for them to thrive.*

3. Look at a pine cone. If we count the number of spirals that are formed in one direction, and count them again in the reverse direction, both results will probably be Fibonacci numbers.

4. Ask your students why they think some flowers or pine cones do not have the usual number of petals or spirals. Just as some humans and animals may be born looking somewhat different from most, flowers and other plants, too, may have a so-called imperfection. This does not make them any less beautiful, however; it may, in fact, make them more interesting than most.

Stefanie Mandelbaum and Jacqueline S. Guttman

For Further Reference

"Historical Topics for the Mathematical Classroom", thirty-first yearbook, National Council of Teachers of Mathematics, Washington, D.C., 1969

Ian Stewart, *Nature's Numbers,* Harper Collins, 1995

Newman/Boles, *Universal Patterns*, Pythagorean Press, Bradford, Massachusetts, 1992

Glossary

Collage a work of art produced by pasting various materials on a single surface.

Descend to have a particular person among one's ancestors, as in "to be descended from..."

Ellipse a plane curve such that the sums of the distances of each point on its periphery from 2 fixed points, (the foci), are equal.

Extreme farthest removed from the average or normal; in math, the first and last terms in a proportion.

Golden exceptionally valuable or advantageous; having the color of gold.

Golden mean the perfect moderate course or position that avoids extremes; the happy medium

Horace (65-8 B.C.) Roman poet and satirist.

Leonardo da Vinci (1452-1519) Italian painter, sculptor, architect, mathematician, musician, and scientist.

Mean in math, (1) average, (2) the second or third term in a proportion.

Moderation avoidance of extremes.

Mondrian (1872-1944) Dutch abstract painter.

Philosophy the rational investigation of the truths and principles of being, knowledge, or conduct.

Proportion comparison; in math, a relation of four numbers such that the first divided by the second is equal to the third divided by the fourth, e.g. 2:1 = 4:2.

Rembrandt (1606-69) Dutch painter

Rubbing a reproduction of a sculptured surface made by laying paper over it and then using pressure and movement with some marking substance, such as pencil or chalk.

Sequence arrangement; in math, a set whose elements have an order similar to that of the positive integers: 1,2,3,4,5....

Seurat (1859-91) French painter.

Veronese (1528-88) (Paulo Calgary) Italian painter from Verona.

Sketch of Funerary Relief in Ostia (golden rectangle)

Sketch of Colosseum

Homework Page

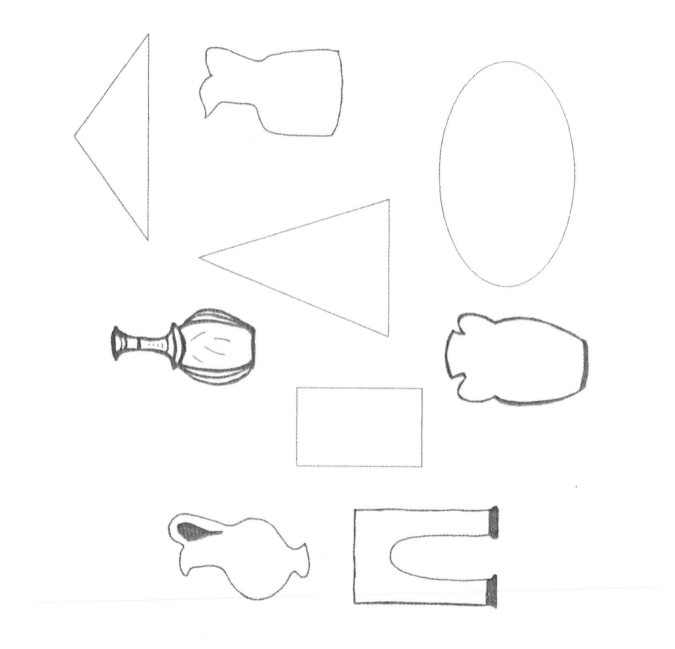

Golden Shape for Art Activity (with previous two pages)

Chapter 6
PYTHAGOREAN RATIOS
IN ART AND MUSIC

At a glance...

Time: Three one-hour sessions or one three-hour session

Activity

After reviewing squares, square roots, and rational and irrational numbers, students learn about the Pythagorean philosophy of harmony and balance in all things. Balance is further explored through the construction of mobiles, the components of which also mirror the musical intervals delineated by Pythagoras.

Mathematical Concepts

- ♦ Rational and irrational numbers
- ♦ Direct and inverse proportions
- ♦ Fractions, decimals, percents, ratios

Pre-planning

Read through the entire lesson ahead of time in order to know where it's going. Decide whether you prefer to divide the lesson into three 60-minute sessions or take an entire morning or afternoon to complete it. Transparencies should be prepared in advance.

Materials/Equipment

1. Plastic straws in four colors (e.g. striped, blue, pink and white). Four straws per student, one of each color. These must be pre-cut and pierced with a needle, pin or sharp pencil point by the teacher, as shown on our diagram. (Older students may be able to do this with a compass point.)
2. Pasta of various shapes, 15 pieces per student. Possible varieties: elbow macaroni, ziti, racchete, rigate, gemelle, farfalle (bow ties), twists.
3. Fishing line, 7 pieces per student, each 12" long
4. Scissors, one pair per student, if possible
5. Calculators, 1 per student, if possible
6. Overhead projector

Step 1 - Math Introduction

Review squares and square roots and rational and irrational numbers.

A number is called rational if it can be expressed as a fraction.

Reinforce: Finding square roots is the opposite of squaring, just as multiplication is the opposite of division, and addition is the opposite of subtraction. Example: The square root of $49 = 7$ because $7^2 = 49$.

Discuss numbers that do not have rational square roots.

Find the square root of 5 on the calculator. It is *not* a rational number because it cannot be written as a fraction. In fact, the calculator gives only an approximate answer. The answer is a decimal that never ends and does not repeat itself. Strangely, when we use a calculator to find the square root of 5, different calculators show slightly different answers. This is because they approximate differently. Our two calculators tell us that the square root of $5 = 2.236067978$, or 2.236067977.

Ask: What happens when you try to square (multiply by itself) the number your calculator told you was the square root of 5? *The answer is not 5!*

Ask: Why does this happen? Is the calculator broken?

The number is not 5 because the calculator was approximating when it found the square root. Our calculators give the answer as 5.000000002 and 4.9999996!

Step 2 - Introduce Pythagoras and his followers.

Pythagoras (pih-THA-gor-us) was born around 570 B.C. on the Greek island of Samos, in the Aegean Sea. He had a great hunger for knowledge, so in order to learn from the best teachers, he traveled very extensively. Pythagoras studied the wisdom of the Chaldeans (chawl-DEE-ans) in Babylon, with Thales (THAL-ees) at Miletus (mil-EE-tis), and is thought to have visited Persia to study the teachings of Zoroaster (zaw-ro-AS-ter). He spent 22 years in Egypt alone because it was an important center of learning. Upon his return to Samos, Pythagoras opened his own school. Unfortunately, the violent and tyrannical government of Polycrates (pah-li-CRAY-tees) caused him to leave his home again at the age of 40 (old for those days). He went to Croton, in southern Italy, where he became involved in local government and made speeches about the "proper" way (according to him) to live one's life.

Pythagoras believed that "Number" was at the root of all things, a universal law as real as light or sound. This philosophy changed people's view of mathematics from something apart from "real" life to the basis for all life. Just as Pythagoras had traveled far and wide to learn from great teachers, many people came to study with him at his school in Samos. Like him, these disciples, called Pythagoreans (pih-tha-gor-EE-ns), believed that the essence of all things was "Number." They believed that there were harmonious proportions among all things, for example between the notes in a musical scale or the distance between the planets and the stars. They also believed that the notes of the musical scale corresponded to the colors of the visible light spectrum.

In the course of their mathematical explorations, the Pythagoreans encountered so-called irrational numbers. Up until that time they had believed that all numbers were rational. When they first became aware of irrational numbers they were very uneasy. After all, they reasoned, if numbers could be irrational, this meant that some proportions in the world, even the world itself, must also be irrational. They were so upset by this discovery that they took a vow never to disclose this fact to the rest of the population! However, one disciple did reveal the theory of irrational numbers to those unworthy to receive it (the general public). He was so hated by the rest of the Pythagoreans that they expelled him from their group and even constructed a symbolic tomb for him, as if he had died. According to legend, the gods were so angry that he perished at sea.

In order to investigate the proportions between the notes of a scale, the Pythagoreans experimented with a monochord, a one-stringed instrument with a movable bridge. (See directions for building a monochord in "Extensions of the Lesson.") This can be demonstrated as follows, using either a monochord or other stringed instrument such as a violin with a tape measure taped to its neck.

♦ Pluck the open string with your forefinger so the students can hear the pitch.

♦ Press your forefinger down at the halfway mark on the ruler and pluck the string with your other forefinger. The string is vibrating at twice the original frequency, making the pitch an octave (8 notes) higher.

♦ If you press the string two thirds of the way down you will hear the note a fifth above the original when you pluck the string.

♦ If you press three quarters of the way down the string and then pluck, you will hear the note which is a fourth above the original.

♦ There is an inverse proportion between the length of the string being plucked and the frequency of the vibrations: as the length of the string goes down (gets shorter), the frequency of the vibrations, or the pitch, goes up (gets higher).

Inform the students that at the next session they will do an art project based on the Pythagorean idea of harmonious proportions and balance.

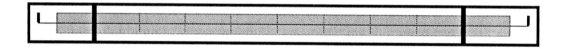

BREAK THE LESSON HERE

PYTHAGOREAN RATIOS, continued...

Step 3 - Making a Mobile

Form groups of six.

(Teacher has pre-cut and pierced through all the straws according to the diagram at the end of the chapter). Distribute pre-cut and pierced straws to students. The straws need not be the colors specified. Each student will have:

one full-size <u>striped</u> straw with 3 pairs of holes: 1 pair at each end and 1 in the middle.

one 3/4 size <u>blue</u> straw with 3 pairs of holes: 1 pair at each end and 1 between them.

one 2/3 size <u>pink</u> straw with 3 pairs of holes: 1 pair at each end and 1 between them.

one 1/2 size <u>white</u> straw with 3 pairs of holes: 1 pair at each end and 1 between them.

Distribute fishing line to students: 7 pieces per student, each piece pre-cut to 12".

Explain that the length of the straws corresponds to the length of the string on the monochord or other stringed instruments.

The <u>striped</u> straw represents the open string.

The <u>blue</u> straw is 3/4 the length of the striped one and corresponds to the fourth note of the scale, which is heard when the plucked string is depressed 3/4 of the way down.

The <u>pink</u> straw is 2/3 of the striped one and corresponds to the fifth note of the scale, which is heard when the plucked string is depressed 2/3 of the way down.

The <u>white</u> straw is 1/2 the length of the striped one and corresponds to the octave above, which is heard when the plucked original string is depressed 1/2 of the way down.

For all of the following, the teacher should demonstrate by creating a mobile step by step along with the students.

Begin construction of the mobile.

- Thread a piece of fishing line through one pair of holes in the striped straw, as shown, and tie it at one end with a double knot so it does not slip through.

- Do the same for the other two pairs of holes. (The center line is to be used to hang up the

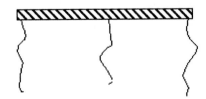

91

mobile.)

- Using the line at <u>one end</u> of the striped straw, thread it through the middle pair of holes in the blue straw. Tie a double knot in this line, 1 inch down. The blue straw is now dangling 1 inch down from the striped straw.

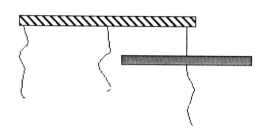

- Repeat this process at the other end of the striped straw, this time attaching the pink straw and tying the knot 1.5 inches down.

- Insert another piece of fishing line at either end of the pink straw and the blue straw as shown. Again, tie each one with a double knot at one end of the line.

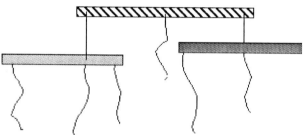

- Thread one end of the string hanging from the pink straw through the middle holes of the white straw, 1.5 inches down. Again, tie a double knot.

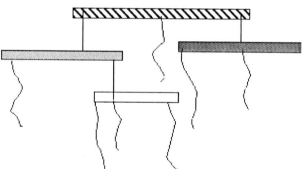

- Insert another piece of fishing line at either end of the white straw. Again, tie each one with a double knot at one end of the line.

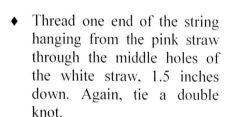

Add the pasta pieces

- In the center of each table put a variety of pasta pieces for the students to choose from. They will need about 15 pieces apiece, depending on the design of their mobiles.

♦ Select one piece and tie it with a double knot part way down one of the lines. Choose another kind of pasta piece for the opposite side and tie it on as well. Continue adding pasta pieces, at various distances down the lines, on all of the end lines and as many of the middle lines as you wish. As you add each piece, hold the mobile up by the middle straw, checking to see how balanced the straws are getting.

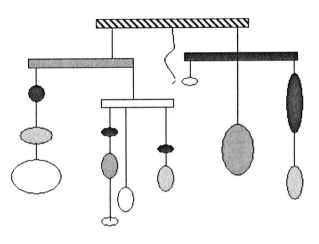

The aim is to get the straws as horizontal as possible.

♦ When the mobile is reasonably balanced, the ends of fishing line are cut off.
♦ Hang the mobile from the line going through the center holes in the striped straw.

Step 4 - Math lesson using students' mobiles.

Select one of the mobiles to work with.

Estimate the proportions used in this mobile.

For example, is the middle string attached about 1/3 of the way over on the pink straw?

What, in the student's estimation, is the percent of the straw that is to the left of the middle string and what percent is to the right?

About how far over is the middle string on the white straw?

What, in the student's estimation, is the fraction of the straw that is to the left of the middle string and what fraction is to the right?

How many pasta pieces are hanging on this mobile?

How many of them are (for example) ziti?

What percent of the total number of pasta pieces are ziti?

What fraction of the total number of pasta pieces are ziti?

Tangents: Related Activities for the Core Curriculum
Language Arts

1. Find the meaning and roots of the word, "philosopher." (philos = love; sophos = wise) What other words are related to it? What do they mean? (ex. philodendron, sophomore, sophistry). Read and discuss Plato's *Allegory of the Cave*. How does this relate to the ideas of the Pythagoreans?

2. Have the class research and write about Alexander Calder, famous for his mobiles. What are the meanings of the word mobile? Why might a brand of gasoline be named Mobil? What is the derivation of auto-mobile? Contrast mobile with stabile.

3. Students might write about their typical day, comparing school days to weekends. Do they think the typical Pythagorean day would lead to a feeling of peace and harmony with the universe? (See page 96.)

4. The Pythagoreans associated different Greek gods and muses with each of these numbers: 1-Zeus, 2-Rhea and Erato, 3-Hecate, 4-Hercules, 5-Aphrodite, 6-Thaleia, 7-Athena, 8-Euterpe, 9-Hera, Terpsichore, and Prometheus, 10-Helios and Atlas. Tell the students some stories from Greek mythology and have them write their own myths. What modern-day heroes would they associate with each number? Why?

Art/Language Arts

It is natural that colors and the names for them are an outgrowth of the geographic area in which a society lives. A desert society might distinguish several shades of tan and brown; people living in a region that is dense with foliage might have distinct words for many shades of green. Similarly, Eskimos have a wide variety of words to describe different types of snow. What is the terrain like in your school's neighborhood? Can the students think of some element that exists in many varieties? For example, in an urban environment, how many different types and colors of sidewalks and streets can be found? Have the class find existing and original words to describe them, e.g. asphalt, cement, concrete, tar, speckled, potholed, bubbly, stone, cobblestone.

Social Studies

1. Myths and Legends

The stories about Pythagoras and his followers contain portions that are probably myths or legends, as opposed to facts. Because these things happened so long ago, and also because much that was written down was destroyed, it is sometimes difficult to separate fact from legend. It may not always be a case of someone lying. We know that different witnesses to a crime, for example, may tell different stories. Are they lying? Generally, they have just perceived things differently from one another.

a. Discuss with the class how we tell truth from fiction. Sometimes we must simply trust our senses.

b. Review which parts of Pythagoras' story are true and which are probably legend. We know that he really existed and that he had a wife and son; that Samos is a real place; that he had followers; that they angered the townspeople. It is probably not true that Hippasus met the "doom of the impious", drowning in the sea for having leaked Pythagorean discoveries to the general public, or that Pythagoras was murdered near a bean field.

c. Sometimes, what one person believes to be truth another may, in the absence of tangible proof, regard as myth. One kind of proof would be ruins of a past civilization, such as Troy.

People first thought that the city of Troy was a myth that had been created by Homer. However, Schliemann, an anthropologist, believed the story, went to where Homer described Troy as being located, and -- lo and behold -- dug up an ancient city! Have your students find examples of familiar myths and legends: the "tooth fairy"; Paul Bunyan; Davy Crockett; Santa Claus; Icarus; the Amazons; Neptune; the lost civilization of Atlantis; UFO's; tales of ancient visits by extraterrestrials, etc. They may be able to ask at home and bring other mythical stories to class. They could create their own legends to explain a given mystery.

2. Greece

Many vestiges of ancient Greek civilizations can be found in Greece even today. Have the class research them, both architecturally and culturally. If any of the students are of Greek origin, they may be able to bring interesting artifacts to share with their classmates.

3. Anthropology

Discuss the work of anthropologists and the ways in which they research ancient civilizations (studying ruins, relics, artifacts, paintings; reading contemporary eyewitness accounts of events).

History/Philosophy/Math

<u>Pythagoras identified four subjects of study:</u>

Arithmetic = Number in itself

Geometry = Number in space

    Music or harmonics = Number in time

Astronomy = Number in space and time

1. The Pythagorean View of Number
    1. is the **monad**, **unity**, the principle of all things; God, the soul, the beautiful, the good, are each <u>one</u>.
    2. is the **dyad**, **duality**, the beginning of conflict, but also the relation of one thing to another.
    3. is the **triad**, two **extremes** bound together by the **mean term**.

MONAD

DYAD

TRIAD

4. is the **tetrad**, the nature of change and righteousness.

5. is the **pentad**, justice, immortality and light.

6. is the **hexad**, health, reconciliation and peace.

| TETRAD | PENTAD | HEXAD |

7. is the **heptad**, the reaper.

8. is the **octad**, all harmonious and steadfast.

9. is the **ennead**, oneness of mind, the absence of strife.

10. is the **tetraktys** (teh-TRAK-tis) or **decad** symbolizing the **cosmos**. (1+2+3+4 = 10).

| HEPTAD | OCTAD | ENNEAD | DECAD |

They felt that unity (1, the monad) leads to two opposite powers (the dyad) which join to generate life. They perceived this as the pattern of the creation of the world.

To the Pythagoreans, each of the first ten numbers had its own meaning and personality.

Do any of these numbers have special meaning for the students? What are their favorite numbers? Why? Ask the students to write a few sentences about what each number means to them. Alternatively, they can choose one number to write a story or essay about. Using the chart at the end of the chapter, they can also copy and color the Pythagorean symbol for the number they chose and use this for the cover of their story or essay.

2. A Typical Pythagorean Day:

- A solitary morning walk before speaking to anyone
- Discussion
- Exercise (running, wrestling, leaping with lead weights on hands, alternating with body-strengthening exercises)
- Bread and honey for lunch, or honey-comb, avoiding wine
- Receptions for guests and strangers
- Another walk, with 2 or 3 companions
- Bath
- Offerings and sacrifices with incense
- Herbs, raw and boiled, maize, wine (those at the highest level did not drink wine because it was thought to be superfluous for them), bread, fish (occasionally)

- Further libations, then readings
- Pythagoreans always wore white clothing and slept in beds covered in white linen
- Food: recommended millet, no beans, no food causing flatulence or indigestion
- Proper music, a vegetarian diet and exercise lead to harmony of body and psyche. Dancing was accompanied by a lyre because they thought that the pipe was too theatrical and would lead to insolence. They believed that playing the monochord, making geometric constructions, and studying abstract mathematics would lead to intellectual harmony. You might wish to try some simple geometric constructions with your class, using compass and ruler.

3. Additional Biographical Information

In 500 B.C., when Pythagoras was 70, a revolt was led against him by Cylon. It is thought that Cylon may have been rejected from the school. Meeting houses were burned and many Pythagoreans perished in the flames. Most of the survivors went to mainland Greece. Ironically, this led to additional centers being established in Athens, Thebes and Phlious. Pythagoras himself escaped to Caulonia, then Locri, then Tarentum, then Metapontum.

We are not sure how Pythagoras died, but it is speculated that it may have been from self-imposed starvation after mobs destroyed and murdered many Pythagoreans. Another legend says that an angry mob chased Pythagoras as far as a bean field. He could have escaped by running across the field, but, believing that beans were sacred, he stood his ground and was murdered.

Although Pythagoras left no books, the survivors left abstracts, commentaries and written memories. He himself left his commentaries to his daughter Damo because his son Telauges was very young when Pythagoras died.

The Pythagorean way of life and study lasted about ten generations. He had about 300 followers. Plato (428-348 B.C.), one of the most famous Greek philosophers, was one of the later Pythagoreans.

Some interesting facts:

- ♦ The Pythagoreans' math discoveries laid the foundation for the development of Greek geometry.

- ♦ Much of Euclid's Elements (the text which is used as the basis for high school geometry) was derived from Pythagorean discoveries.

- ♦ They knew how to find a "golden cut," divine proportion, golden section. (See previous chapter).

4. Philosophical Questions

This chapter provides a good opportunity to introduce students to the study of philosophy, from its definition in the abstract to developing a philosophy of life. It is not too soon to pose such proverbial philosophical questions as: What is truth? The standard question -- If a tree fell in the forest and no one heard it, would it actually have made a sound? -- is a good place to start. What if no one saw it -- did it still fall? What if an animal saw it? Another question, for more sophisticated children, might be: What is art?

Science

1. What do the students feel would lead to intellectual harmony (a state in which the mind is totally accepting and at peace with its perceptions of reality)? How do they feel about vegetarian diets? What do they like to eat? Do different foods affect their moods? (Chocolate and endorphins, for example, or carbohydrates and quick energy).

2. Studies have shown that, after listening to Mozart, students do better in mathematics. Why do they think this is? Is there a connection between music and mathematics? What music do the students prefer? Do they study better with music or in silence? Can they concentrate better after listening to certain kinds of music than others? Class experiments might be tried using various types of music followed by math problems. The results could then be analyzed and evaluated.

Music

INSTRUCTIONS FOR BUILDING A MONOCHORD

Materials:

2"x 4"x 48" lumber (unwarped)

2 screw eyes or tuning pegs

5' piano or harpsichord wire (.010 or .014 - the heavier wire gives a louder sound)

flat wooden ruler - one yard or one meter

2 pieces of strong wood, .5" x .75" x 2" (for bridges).

- ♦ Glue the ruler onto center of the 2x4.

- ♦ Glue or bolt bridges onto the 2x4, one at either end of the ruler.

- ♦ Put the screw eyes (or tuning pegs) in center, near edges of 2x4.

- ♦ Loop piano wire into one screw eye, pass it over both bridges and loop it into the other screw eye.

As with the monochord's divided string, the smaller any (non-electronic) instrument, the higher its pitch. Think of the piccolo compared with the tuba, or a violin versus the string bass. An experiment: collect 8 glass bottles of the same size. (Soda bottles work well.) How much must you fill each in order to play the notes of the scale? The bottles can be struck or blown across to play pitches.

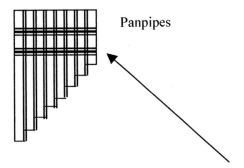

Panpipes

Music/History

1. Research the history of the instruments we use now. Pythagoras had pipes, for example. Each tube played a different note when one blew into it. How have these evolved into the modern-day flute? How has the lyre of Pythagoras' day evolved?

2. Make your own string instruments. Stretch rubber bands across a wooden or cardboard box, or a shoebox. The pitch varies according to how heavy the rubber band is and how tightly it's stretched. Just as with the monochord, if you hold the rubber band halfway across and pluck it, you will hear a tone one octave above the open "string".

Art

1. <u>Color:</u>

Pythagoreans believed that only ten colors existed: white, black, yellow, tawny (a mustard color), pale (off-white), red, green, blue, light blue, grey. This may be because these were the only colors in their world. Their dyes came from nature, e.g. clay, fruits and flowers.

   a. Look around the classroom. What colors do your students see that are not on their list? Which ones from the list are not seen in the room?

   b. Try a class art project using only these colors, e.g. a landscape or scene from their neighborhood. Discuss what other colors they might want to use if they could, e.g. pink, purple, brown, orange.

   c. Mix some of these colors to see what other ones can be created. Examples: red/blue, yellow/blue, red/yellow.

   d. Art/Science: Research how colors are created today. What kinds of dyes are used?

   e. Have a "Pythagorean Day" on which the only colors worn are ones from their list.

   f. Mix natural colors with your class from beets or black walnuts, for example, to illustrate how the people in Pythagoras' time would have made dyes.

2. <u>Mobiles</u> are a wonderful way to learn about balance while illustrating all kinds of subjects: nature (branches with leaves, feathers, dried plants); families and friends (hangers with cut out photos); animals, etc. Found objects such as beads and buttons are also useful and fun.

Stefanie Mandelbaum and Jacqueline S. Guttman

For Further Reference

Kenneth Sylvan Guthrie, *The Pythagorean Sourcebook and Library*, compiled and translated by Phanes Press, 1987

Homage to Pythagoras, Lindisfarne Letter, Lindisfarne Corresponding Members Conference, Crestone, Colorado, 1981

Anne and Christopher Moorey, *Making Mobiles*, Watson-Guptill Publications, New York, 1966

Muriel Mandell and Robert E. Wood, *Make Your Own Musical Instruments*, Sterling Publishing Co., Inc., New York, 1957

Reinhold Banek and Jon Scoville, *Sound Designs, A Handbook of Musical Instrument Building*, Ten Speed Press, Berkeley, California, 1980

Mike Jackson, *Making Music*, Angus and Robertson, Australia, 1992

Glossary

Anthropologist one who specializes in the science that deals with the origins and physical, cultural, and social development of a given group of people.

Artifact any object made by man, especially with a view to subsequent use (from *arte factum*- something made with skill).

Bridge (on a musical instrument) a thin, fixed wedge or support raising the strings of a musical instrument above the sounding board.

Cosmos the world or universe regarded as an orderly, harmonious system.

Disciples people who are adherents of the doctrines of another.

Harmonious proportions congenial, pleasing relationships.

Homer 8th Century Greek epic poet, reputedly the author of the Iliad and the Odyssey.

Irrational numbers numbers which are not equivalent to any fractions.

Lyre (LIAR) A musical instrument of ancient Greece consisting of a soundbox, typically made from a turtle shell, with two curved arms connected by a yoke from which strings are stretched to the body.

Mobile (MO-beel) a piece of sculpture having delicately balanced units constructed of rods and sheets of metal or other material, suspended in midair by wire or twine so that the individual parts can move independently.

Monochord an acoustical instrument dating from antiquity, consisting of an oblong wooden sounding box, usually with a single string, used for the mathematical determination of musical intervals.

Philosopher a person who offers his/her views or theories on profound questions in ethics, metaphysics, logic and other related fields.

Relic a surviving memorial of something past.

Schleimann (SHLEE-man) (1822-90) German archaeologist and expert on Homer.

Stabile (STAY-beel) a piece of sculpture having immobile units constructed of sheet metal or other material and attached to fixed supports.

Troy an ancient ruined city in Northwest Asia Minor.

Tyrannical unjustly cruel, harsh or severe.

Universal pertaining to the whole.

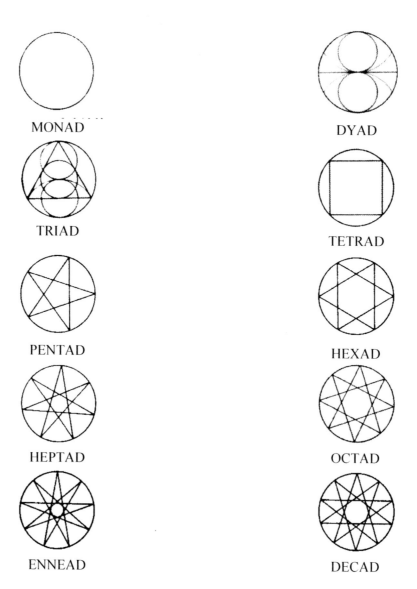

MONAD

DYAD

TRIAD

TETRAD

PENTAD

HEXAD

HEPTAD

OCTAD

ENNEAD

DECAD

Striped straw

Blue straw

Pink straw

White straw

Monochord

Chapter 7
FIGURATIVE FRACTIONS

At a glance...

Time: Two 1-hour sessions plus an assignment.

Activity

An investigation of the proportions of Native American totem poles leads to the application of fractions and percentages in the design and construction of students' own. Discussion takes place about the history, uses and significance of totem poles in Native American societies.

Mathematical Concepts

- ♦ Addition, subtraction, multiplication and division of whole numbers and fractions
- ♦ Reducing fractions
- ♦ Least common denominator
- ♦ Ratios and proportions
- ♦ Estimating fractions and percents

Pre-planning

- ♦ Read through the entire lesson ahead of time in order to know where it's going. If necessary, modify the lesson (e.g. the specific fractions) according to the abilities of your class.
- ♦ Draw a rectangle 40 boxes long by 23 boxes wide, centered on 8 ½"x11" graph paper.
- ♦ Draw a rectangle 40 boxes long by 12 boxes wide, centered on 8 ½"x11" graph paper. (For optional large totem pole)

Transparencies and photocopies should be prepared in advance.

Materials/Equipment

1. Transparencies of totem pole examples made from our black line masters
2. Photocopies of diagrams of totem poles (one per student)
3. Photocopies of the rectangle 23 boxes wide on graph paper (one per student)
   Note: Photocopy the rectangle onto the heaviest possible paper.
4. Pencils with erasers
5. Notebook or scrap paper for calculations
6. Cardboard or other stiff material, 6"x11" (one per student)

7. Colored pencils, crayons, pastels, markers or paint and brushes
8. Rulers
9. Glue or staplers
10. Overhead projector

Step 1 - Totem Introduction

Show the overhead transparency of *Totem II.*

This is an ink drawing by artist Stefanie Mandelbaum called *Totem II.* It is based on Haida totem poles.

Traditional totem poles are 3-dimensional, usually carved out of cedar wood, made from a single tree trunk from which the branches have been stripped off. They are then split in half lengthwise, leaving the back of the totem flat. Protruding parts, such as wings, are attached with pegs.

The largest totem pole in the 19th century was 80 feet tall; in the 20th century they rose as high as 173 feet. Totem poles generally last about 60 to 80 years; they are destroyed primarily by weather and insects. After the poles are carved, they are painted. Before Europeans came to America, paints were often ground from minerals mixed with salmon eggs. Today, however, commercial paints are used.

It takes many people to raise a finished pole, over 100 to carry it and many more to dig the hole that holds the base. The pole is leaned into the hole with ropes tied around it; many people then pull on the ropes to get it upright. Raising the totem pole was cause for special ceremonies and a great feast, called a *potlatch*. Carving and raising a totem pole and hosting the potlatch was very expensive, and could only be afforded by tribal chiefs.

Some fine examples of totem poles can be seen in the Pacific Northwest, from the *Haida* (HY-da), *Tlingit* (TLING-it) and *Kwakuitl* (kwa-KYU-tl) tribes. They serve many different purposes: as an entry portal (with a hole in the base for the doorway); as a memorial to help remember a special event; for protection from evil; to welcome visitors (like a welcome mat). Many depict legends or myths, just as European religious paintings or sculpture often illustrated bible stories. In both cases, these visual stories were created because the people could not read. In some cultures, there was no written language.

Point out: Totem poles generally contain bilateral symmetry -- two halves that reflect one another.

Ask: What do you see in this room that also has bilateral symmetry?

Our faces and bodies, possibly a window or chalkboard

Point out: Totem poles look like many separate sculptures of varying heights that have been stacked on top of each other. The pole is divided unevenly. Take note of the relative size, or proportions, of different sections.

Show the transparency of Mandelbaum's *Temple Totem.*

Temple Totem is made of abstract forms but is still influenced by traditional Haida totems. Instead of wood, Mandelbaum makes her totems from modern materials such as styrofoam, steel, cement and fiberglass. Like the Haida totems, their sections are also divided unequally. When the artist made this totem, she was thinking about the way totems are sometimes thought to provide protection, which is one reason she made it out of styrofoam. In our culture, styrofoam is often used to protect things like computers, VCR's and other electronic equipment. Although it is abstract, this totem is meant to express the same kinds of feelings as traditional totem poles.

.Ask: Why do you think this was called *Temple Totem?*

The bottom section resembles the entrance to an ancient temple.

Show the transparency of Mandelbaum's *Complementary Totem.*

This piece is also made of styrofoam, steel and cement and divided into four sections, but the sections are in different ratios than the ones in *Temple Totem.* It got its name because it is painted in pink and green, which are opposite each other on the color wheel—that is, complementary colors.

Many other contemporary artists have been influenced by traditional totem poles. Among them are David Smith, Dorothy Dehner, and Louise Nevelson, sculptors whose totems are of steel, bronze and wood, and painter Jackson Pollack.

Step 2 Math Introduction

Distribute copies of the drawing of *Temple Totem.*

Point out: *Temple Totem* is divided into 4 unequal parts that are identified as A, B, C and D.

The height of each section is indicated on the side.

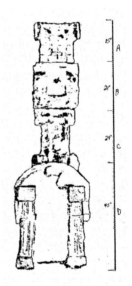

Suggested questions:

What is the total height of the pole? *105", or 8' 9"*

Which section's height is 1/5 of the whole? *B*

Which section's height is 1/7 of the whole? *A*

What fraction of the whole is the height of section D? *3/7*

Section C? *8/35*

Add the four fractions together, using the least common denominator of 35: *1/7+1/5+8/35+3/7=5/35+7/35+8/35+15/35=35/35=1, or the entire height*

Which section is 1/3 of the height of another section? *A is 1/3 of D*

Another way of looking at this is D/3=A or 3/7÷3 =1/7. This shows that (3/7) x (1/3) = (1/7). (Note: Percents can be substituted for fractions.)

Repeat the process with *Totem II*

Totem II is 14 feet long and divided into six parts.

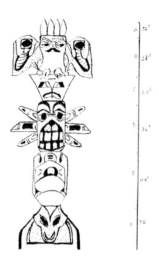

Suggested questions:

How many inches equal 14'? *168"*

The sections measure, from top to bottom: A=12" B=28" C=20" D=36" E=40" F=32"

Fractions of the whole are: A=1/14, B=1/6, C=5/42, D=9/42, E= 5/21, F=4/21

What fraction of the whole is section A? *12/168 = 1/14*

What fraction of the whole is section B? *28/168=1/6*

What fraction of the whole is section C? *20/168=5/42*

What fraction of the whole is section D? *36/168=9/42*

 A+B+C+D+E = 136/168 = 17/21

What is left when you subtract 17/21 from 21/21?

4/21, which is the size of section F

4/21 equals "what?" over 168? *32*

Step 3: Preparation for art-math activity

Distribute vertical rectangles on heavy graph paper as photocopied.

The rectangle is 40 boxes high.

Explain to the class that they can divide the rectangle into various proportions.

Suggested questions:

How many boxes down from the top of the rectangle would be 1/8 of the way down?

5 boxes

Students draw a horizontal line 5 boxes down.

How many boxes are left? *35*

What is 1/5 of what is left? *7*

Students draw their next horizontal line 7 boxes down from the first line.

How many boxes have we used so far? *12*

How many are left? *28*

What fraction of the whole is left? *28/40 = 7/10*

What is 1/7 of 28? *4*

Students draw the next line 4 boxes below the previous one, leaving 24 boxes.

What is 3/8 of 24? *9*

Draw the next line 9 boxes lower.

How many boxes are left? *15*

What is 1/5 of 15? *3*

Draw the next line 3 boxes lower.

How many boxes are left now? *12*

What fraction (or percent) of the whole is each part?

A = 5/40 = 1/8

B = 7/40

C = 4/40 = 1/10

D = 9/40

E = 3/40

F = 12/40 = 3/10

Add the fractions. They should add up to 40/40 = 1.

Note: One of the most difficult concepts for children to absorb is that of the *least common denominator*. This exercise helps them visualize the common denominator, which in this case is the total number of boxes in the height of the rectangle.

BREAK THE LESSON HERE IF YOU WISH

Homework (or classroom) assignment: As preparation for making their individual totems, send each student home with the photocopied rectangle (on heavy paper, if possible) and the homework sheet.

FIGURATIVE FRACTIONS continued...

Step 4: Students make their own totem poles

Distribute the 6"x11" cardboard. Have the students take out the assignment that they completed in preparation for this activity.

On the graph paper, have the students fold the paper lengthwise along each vertical edge of the rectangle.

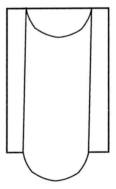

Students draw their own totem poles on the graph paper. They can base their drawings on anything: their friends or families, animals, holidays, superheroes -- and be as abstract as they wish. The possibilities are limitless!

Color the drawings and then glue or staple the margins onto the cardboard so that the drawings bulge out to form a semi-cylinder.

Students take turns showing their works to the class which can estimate what fraction or percent of the whole any given section represents, after which the students "show and tell" each other what proportions they actually chose for their totem poles.

If desired, the totem poles can be tacked up one under the other to form longer poles for display purposes. In this case you might wish to cut off the undecorated tops and bottoms of the poles (the areas above and below the 40 boxes), or ask students to color them as well.

MAKING LARGE TOTEM POLES - GROUP PROJECT

An additional lesson in which teams of students use a grid to create 3 or 4 large totems.

Materials

- 8 ½" x 11" photocopies of narrow rectangles 12 boxes wide x 40 boxes high (1 per student)
- 3 or 4 sheets of 42" x 15" graph paper made up of 1" squares (1 sheet per group), available in art supply stores
- 3 or 4 cardboard sheets, 42" x 13" (1 sheet per group)

Form small groups of six students to a team. Distribute one narrow rectangle to each student.

Divide each rectangle into 6 sections as described in step 4 above.

Each student in the group is then assigned one section in which to draw. Students draw *only* within the section that has been assigned to them. When the drawings are finished, the students cut out the section on which they have drawn.

Each group then combines and rearranges the drawn and cut out sections until the design looks right to them. They number the sections in order on the back, top to bottom. Each student then takes back his or her own section.

Distribute one sheet of 1" square graph paper to each group.

The group draws a rectangle on this graph paper in the same proportions as the original smaller rectangle. Section this off with horizontal lines drawn edge to the edge as was done on the smaller rectangle, keeping the same proportions. (Note: Rectangle should be centered on the sheet.)

The sections are cut apart along the horizontal line. Each student takes back to his/her seat the (small) original cut-out section and the (large) new one that corresponds to it.

Looking at the upper left hand box of the small version, they redraw *just that box* on the larger version. (The new one is 4 times the size of the original.) They then proceed to the next box, continuing until they have finished the entire section.

Color the drawings.

On the floor if necessary, arrange the sections face down in the order decided upon earlier. Tape the sections together on the back.

To display the totem poles, fold them vertically along the edges of the rectangle and attach the edges to the cardboard as before. They will lean against a wall without much assistance.

The small square [ ] is to the large square [ ] as

the drawing in the small square [ ] is to the drawing in the large square [ ] as 1:4.

> **Note:** The above lessons can be adapted for students learning percents. For example, instead of determining what fraction of the whole a given section represents, students can determine what percent of the whole is represented.

Tangents: Related Activities for the Core Curriculum
Social Studies
1. Study more about the Northwest Coast peoples, including the history of what happened to them when the Europeans came. Among other things, they were forbidden to make totem poles because they were considered heretical (idol worship). Although they stopped making them for many years, they eventually started again and continue to this day.
2. Research other cultures that use totems, such as the Maori in New Zealand.
3. Discuss symbols of protection in own cultures and families. What do we do to "protect" ourselves from danger? (e.g. "good luck" charms we carry or wear, religious symbols on dashboards, etc.)

Science
1. Study how paints and dyes were made by native peoples and compare them with modern paint manufacture. Which are safer? Use this as an opportunity to teach about the dangers of lead poisoning from old house paint.
2. Trees: Investigate why cedar was usually used to make totem poles. What else is cedar used for? What are its properties?
   - Research the trees that grow in your vicinity, perhaps through a local field trip. Are there cedar trees nearby? What other kinds of trees might also make good totem poles? Why?
   - Is it environmentally sound to use trees to make totem poles? What do trees do for us?
   - What is the attitude of the Haida and other Native Americans toward trees and the environment in general?
3. Styrofoam is a very controversial product. What is it made from? Discuss its advantages and disadvantages.

Language Arts
1. Have the students read aloud the legends of the Northwest Coast peoples.
2. Ask them to create their own oral or written legends based on the ones that are depicted by their totem poles.

Art
1. Learn more about the American artists David Smith, Dorothy Dehner, Louise Nevelson and Jackson Pollack, especially their totem sculptures and paintings.
2. The people who made totem poles also produced other kinds of beautiful art and artifacts. Learn more about their weaving and pottery.

Stefanie Mandelbaum and Jacqueline S. Guttman

For Further Reference

David Campbell, ed., *Native American Art and Folklore, a Cultural Celebration*, Crescent Books, New York, 1993

Bill Holm, *Spirit and Ancestor*, University of Washington Press, Seattle, 1987

Lewis Spence, *The Illustrated Guide to Native American Myths and Legends*, Longmeadow Press, Stamford, 1993

Rosalind E. Krauss, *The Sculpture of David Smith*, MIT Press, Cambridge, 1971

Dawns and Dusks: *Louise Nevelson*, Conversations with Diana MacKown, Charles Scribner's Sons, New York, 1976

Glossary

Abstract art which is nonrepresentational and may not be recognizable as an object, figure or scene

Bilateral Symmetry (by-LAH-ter-ul) having, or involving, two sides that mirror each other on opposite sides of an axis

Complementary colors colors opposite each other on the color wheel: red and green, blue and orange, violet and yellow, for example

Grid a basic system of reference lines, consisting of straight lines intersecting at right angles

Potlatch a ceremonial festival at which gifts are bestowed on the guests

Proportion a relation among four amounts in which the first divided by the second is equal to the third divided by the fourth. For example: 5:10 as 1:2, or 5/10 =1/2

Ratio the relation between two similar quantities with respect to the number of times that the first contains the second

Totem a natural object or living being, such as an animal or bird, which is the symbol of a clan, family or group

Totem pole a post carved and painted with totemic figures

Vertical upright

Native Peoples

Haida (HY-da) a Native American group of people inhabiting the Queen Charlotte Islands in British Columbia and Prince of Wales Island in Alaska

Kwakuitl (kwa-KYU-tl) Native American people of Vancouver Island and the British Columbian coast

Tlingit (TLING-it) Native American people inhabiting the coast of southern Alaska and northern British Columbia

GETTING READY TO MAKE A TOTEM
(Homework Sheet)

You have been given a rectangle that is now 40 boxes high by 23 boxes wide.

1. Divide the rectangle into 4 to 8 sections, top to bottom. Make the sections any size you wish.

2. Label the sections A, B, C, D, etc.

3. Count the vertical boxes in each section. How many boxes high is each section?

4. Figure out what fraction of the whole rectangle each section is, using the number of boxes divided by 40 (the number of boxes in the whole rectangle). For example, if section A is 10 boxes high, divide 10 by 40 (10/40).

5. If possible, reduce the fraction. (10/40) = (1/4)

6. Add up all the fractions to see if they equal 40/40, or 1.

7. Bring this with you to class. You are now ready to make your own totem pole.

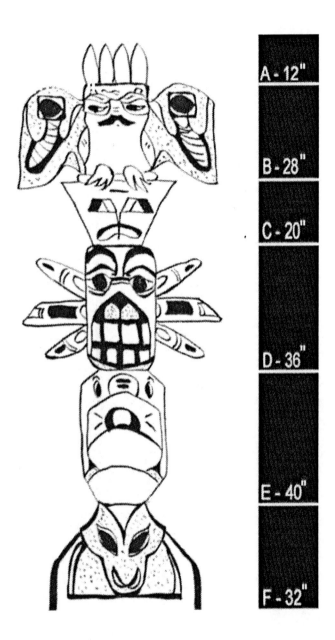

Totem II

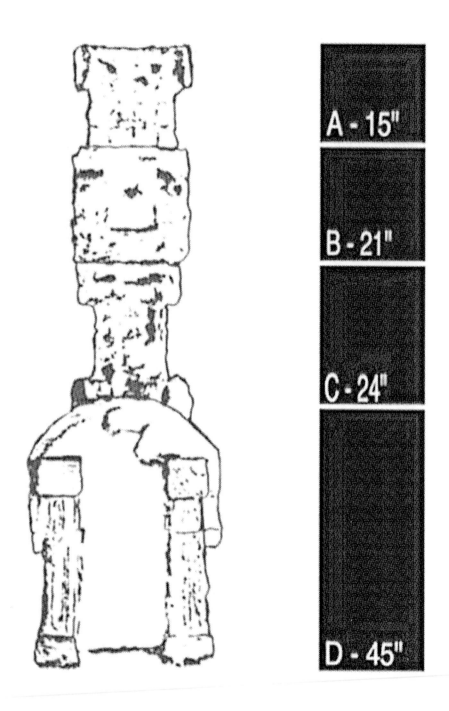

Temple Totem

Complementary Totem

Temple Totem

Appendix

Numerous websites are available for examples of artwork to supplement each chapter. Some excellent ones are listed below.

1. www.asianart.com
   (Mystical Mandalas)
2. Brooklyn Museum of Art, Brooklyn, NY
   www.brooklynart.org
   (2-D in a 3-D Universe, Collages and Composites, The Golden Mean, Figurative Fractions)
3. Coghlan Art Studio and Gallery, British Columbia, Canada
   www.coghlanart.com
   (Figurative Fractions)
4. The International Museum of Collage, Assemblage, and Construction
   Cuernavaca, Mexico
   http://ipdg.org/museum/collage/
   (Collages and Composites)
5. The Math Forum
   http://forum.swarthmore.edu
   (Kitchen Floors and the Alhambra Palace)
6. The Metropolitan Museum of Art, New York, NY
   www.metmuseum.org
   (2-D in a 3-D Universe, Kitchen Floors and the Alhambra Palace, Collages and Composites, The Golden Mean)
7. Museum of Anthropology, University of British Columbia, Canada
   www.moa.ubc.ca
   (Figurative Fractions)
8. The National Gallery of Art, Washington, D.C.
   www.nga.gov
   (2-D in a 3-D Universe, Kitchen Floors and the Alhambra, The Golden Mean, Pythagorean Ratios in Art and Music)
9. The State Hermitage Museum, St. Petersburg, Russia
   www.hermitagemuseum.org
   (2-D in a 3-D Universe, Mystical Mandalas, The Golden Mean)
10. WebMuseum, Paris, France
    http://www.ibiblio.org/wm/
    (2-D in a 3-D Universe, Collages and Composites, Pythagorean Ratios in Art and Music)

About the Authors

Stefanie Mandelbaum, an artist/mathematician, is an adjunct assistant professor at Rider University in Lawrenceville, NJ, where she has taught in the Education, Fine Arts, and Management Science departments. Her courses have ranged from *Teaching Mathematics: Nursery through Grade 8* and *Introduction to Quantitative Methods* to *Art and Society* and *Nineteenth* and *Twentieth Century Art*. In addition, she has taught both math and art on all levels in school and community settings and has worked extensively with students suffering from math anxiety, learning disabilities and attention-deficit disorder. Mandelbaum holds a BS in mathematics from Queens College (CUNY), a Master of Arts in Teaching (mathematics) from Montclair University and an MFA (sculpture) from Pratt Institute. The lessons in *ARThematics Plus: Integrated Projects in Math, Art and Beyond* are based on over 300 student and teacher workshops she has given since 1994 throughout New Jersey and the tri-state area.

Jacqueline S. Guttman is founding partner at ArtService Associates, an arts management consulting organization with an emphasis on arts education and outreach. A former flutist and teacher of music in grades K-12, she has written numerous teacher supplements for performances and arts workshops in the public schools which focus on integrating the arts with all subject areas. For several years she was an adjunct assistant professor in the Performing Arts Administration program at New York University, where she also supervised student internships and was acting director. Guttman holds a BS in music education from Potsdam College of the State University of New York, and Masters Degrees in both music education and arts administration from New York University. Not incidentally, she was a math-anxious student throughout her school career.

We dedicate this book to our husbands, David Mandelbaum and Howard Guttman, who patiently and lovingly supported us through five years of meetings, telephone conversations, e-mails, revisions and more revisions – and *more* revisions.

Printed in the United States
28096LVS00001B/23-98